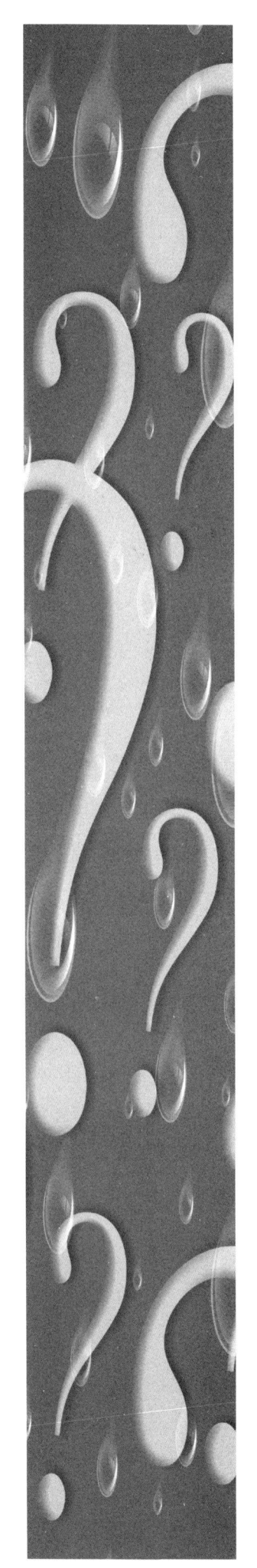

OPERATOR CERTIFICATION STUDY GUIDE

A Guide to Preparing for Water Treatment
and Distribution Operator Certification Exams

Fifth Edition

Written by John Giorgi
and prepared by
the Association of Boards of Certification

Disclaimer

Although this study guide has been extensively reviewed for accuracy, there may be an occasion to dispute an answer, either factually or in the interpretation of the question. Both AWWA and ABC have made every effort to correct or eliminate any questions that may be confusing or ambiguous. If you do find a question that you feel is confusing or incorrect, please contact the AWWA Publishing Group.

Additionally, it is important to understand the purpose of this study guide. It is not a magic bullet to certification. It is intended to provide the operator with an understanding of the types of question he or she will be facing on the certification exam and with the areas of knowledge that will be covered. AWWA highly recommends that you make use of the additional resources listed at the end of this guide in preparing for your exam.

Copyright © 1993, 2003 American Water Works Association.
All rights reserved.
Printed in the United States of America.
ISBN 158321-287-6

American Water Works Association

6666 West Quincy Avenue
Denver, Colorado 80235-3098
303.794.7711

TABLE OF CONTENTS

Preface, v
Acknowledgments, ix

Water Treatment

Evaluate Characteristics of Source Water, 3
Monitor, Evaluate, and Adjust Treatment Processes, 8
Laboratory Analysis, 14
Operate Equipment, 19
Evaluate and Maintain Equipment, 24
Safety and Emergency Preparedness, 29
Perform Administrative Duties, 34
Additional Sample Questions, 39
Math For More Practice, 58

Water Treatment Answers

Evaluate Characteristics of Source Water, 71
Monitor, Evaluate, and Adjust Treatment Processes, 75
Laboratory Analysis, 80
Operate Equipment, 84
Evaluate and Maintain Equipment, 88
Safety and Emergency Preparedness, 92
Perform Administrative Duties, 96
Additional Sample Questions, 100
Math For More Practice, 115

Distribution

System Design, 131
Monitor Water Quality, 136
Install Units, 142
Operate and Maintain Equipment, 147
Safety, 153
Perform Administrative Duties, 158
Additional Sample Questions, 165
Math For More Practice, 184

Distribution Answers

System Design, 197
Monitor Water Quality, 201
Install Units, 205
Operate and Maintain Equipment, 209
Safety, 213
Perform Administrative Duties, 217
Additional Sample Questions, 221
Math For More Practice, 233

Appendix A, 245

Additional Resources, 251

This work is dedicated to my children, Sara, Stephanie, and Steve, and to my mother, Thelma Giorgi.

PREFACE

The American Water Works Association (AWWA) and the Association of Boards of Certification (ABC) worked together to develop this guide based on the ABC Need-to-Know Criteria. This guide is intended to give operators practice answering questions that are similar in format and content to the questions that appear on certification exams. The questions in this study guide are not the same questions that will appear on the certification exam. However, the questions allow operators to experience the types of questions that may be on a certification exam. If you have difficulty answering any of the questions in the study guide, you should consult the reference source provided following the answer to the question.

American Water Works Association

The American Water Works Association (AWWA) is an international nonprofit scientific and educational society dedicated to the improvement of drinking water quality and supply. AWWA is defined by six core competencies, through which we communicate and interact with all of our audiences. These competencies relate to advocacy, communications, conferences, education and training, and science and technology. Together, the competencies distinguish AWWA as the authoritative resource for knowledge, information, and advocacy to improve the quality and supply of drinking water in North America and beyond.

Founded in 1881, AWWA is the largest organization of water supply professionals in the world. Its more than 50,000 members represent the full spectrum of the drinking water community: treatment plant operators and managers, scientists, environmentalists, manufacturers, academicians, regulators, and others who hold genuine interest in water supply and public health. Membership includes more than 4,000 utilities that supply water to roughly 180 million people in North America.

Association of Boards of Certification

The Association of Boards of Certification has been assisting states and provinces with environmental certification programs since 1972. ABC's mission it to provide an organization for certifying authorities that will facilitate communication and cooperation. ABC's goal is to help environmental certifying authorities do a better job by sharing information, know-how, and resources. ABC believes this will enable all to better fulfill their responsibilities to ensure the competence of environmental occupations and laboratories. ABC's objectives are to:

- Improve and strengthen certification laws
- Promote certification as a means of ensuring effective operation
- Define and maintain internationally recognized qualifications for certification in established categories
- Promote a uniformity of standards and practices in certification
- Facilitate the transfer of certification between certifying authorities
- Assist newly created certifying authorities in establishing initial policies and procedures

ASSOCIATION OF BOARDS OF CERTIFICATION OPERATOR EDUCATION AND EXPERIENCE REQUIREMENTS

ABC's education and experience requirements for water treatment operators are as follows:

Class I

- High school diploma, general equivalency diploma (GED), or equivalent; and
- One year of acceptable operating experience of a Class I or higher utility.
- No substitution for experience shall be permitted.

Class II

- High school diploma, GED, or equivalent; and
- Three years of acceptable operating experience of a Class I or higher utility.
- A maximum of 675 contact hours, or 68 continuing education units (CEUs), or 68 quarter credits, or 45 semester credits of post high school education in the environmental control field, engineering, or related science may be substituted for one and one-half years of operating experience.

Class III

- High school diploma, GED, or equivalent; and
- 900 contact hours, or 90 CEUs, or 90 quarter credits, or 60 semester credits of post high school education in the environmental control field, engineering, or related science; and
- Four years of acceptable operating experience of a Class II or higher utility, including two years of direct responsible charge.
- A maximum of 900 contact hours, or 90 CEUs, or 90 quarter credits, or 60 semester credits of appropriate post high school education in the environmental control field, engineering, or related science may be substituted for two years of experience; however, the applicant must still have one year of direct responsible charge experience.

- A maximum of one year of direct responsible charge experience in a Class II or higher position may be substituted for 450 contact hours, or 45 CEUs, or 45 quarter credits, or 30 semester credits of post high school education in the environmental control field, engineering, or related science.

Class IV

- High school diploma, GED, or equivalent; and
- 1,800 contact hours, or 180 CEUs, or 180 quarter credits, or 120 semester credits of post high school education in the environmental control field, engineering, or related science; and
- Four years of acceptable operating experience of a Class III or higher utility, including two years of direct responsible charge.
- A maximum of 900 contact hours, or 90 CEUs, or 90 quarter credits, or 60 semester credits of appropriate post high school education in the environmental control field, engineering, or related science may be substituted for two years of experience; however, the applicant must still have one year of direct responsible charge experience.
- A maximum of two years of direct responsible charge experience in a Class III or higher position may be substituted for 900 contact hours, or 90 CEUs, or 90 quarter credits, or 60 semester credits of post high school education in the environmental control field, engineering, or related science.

Substitutions

- Education applied to the operating and direct responsible charge experience requirement shall not be applied to the education requirement.
- Also, operating or direct responsible charge experience applied to the education requirement shall not be applied to the operating or direct responsible charge experience requirement.
- One year of operating or direct responsible charge experience may be substituted for one year of high school, without limit.
- Where applicable, related experience in maintenance, laboratories, other environmental control utility positions, and allied trades such as plumbing, or other certification categories may be substituted for one half of the operating or direct responsible charge experience requirement; however, the applicant for Class III and IV must still have one year of direct responsible charge experience.
- The maximum substitution of education and related experience for operating or direct responsible charge experience shall not exceed 50 percent of the stated operating or direct responsible charge experience requirement.

ACKNOWLEDGMENTS

The following people served as reviewers for this edition of the *Operator Certification Study Guide*. Their assistance is greatly appreciated.

Bruce Baldwin, Superintendent, Seekonk Water District

Richard Bond, Colorado Springs Utilities

Pat Cook, Treatment Specialist, DEQ-Water Division

Don Davis, O'Brien & Gere Operations, Inc.

Ernest Earn, Environmental Facilities Assessment Coordinator, Environmental Protection Division

Louis Gialanella, Director of Water Quality, Brick Township Municipal Utilities

Don Jackson, Pipeline Supply Company

Dorian Jefferson

Jess Jones, OTCO

Ken Kerri, CSUS Office of Water Programs

Dan Laprade, Circuit Rider—Drinking Water Program, DEP Springfield

Anne Lynam, Dept. of Natural Resources

Andrew Maguire, E.L. Smith WTP

Andrew Maquire, EPCOR Water Services Inc.

Cheryal McDaniel, Kimzey Regional Water Distribution District

John McEncroe, Water Treatment Superintendent, City of Golden

Bob Morphis, City of Aurora

Martin Nutt, Drinking Water Advisory & Op Licensing

Mike Ranger, Superintendent of Treated Water Operations, Denver Water

Paul Riendeau, Education Coordinator, NEWWA

Thomas Rothermich, City of St. Louis—Water Division

David Secor, Ohio EPA, SWDO

Jon Strutzel, DHS, Certification Unit, MS 92

Steven Sullivan, DPW-Water and Sewer Division

Mike Thomason, Arkansas Environmental Academy

Steve Wear, Conway County Regional Water Distribution District

John Giorgi would like to thank his water treatment associates at El Dorado Irrigation District in Placerville, CA: Eric Cosens, Margie Lopez-Read, Billy Munn, Patty Kline, John Bowen, Matt Boring, Johnny Wilson, Radenko Odzakovic, Tod Granicher, Susan Durham, Kurt Mikkola, and Gary Meyers. Also his students and work associates: Robin Peck, Zol Whitman, Jake Maker, Matt Heape, Daryl Herren and Darren Kitzmiller.

WATER TREATMENT OPERATOR
CERTIFICATION EXAMS

EVALUATE CHARACTERISTICS OF SOURCE WATER

	Certification Level			
	I	II	III	IV
Evaluate bacteriological characteristics of source water	X	X	X	X
Evaluate biological characteristics of source water	X	X	X	X
Evaluate chemical characteristics of source water	X	X	X	X
Evaluate physical characteristics of source water	X	X	X	X

Suggestions for Study:

- Knowledge of normal characteristics of water
- Knowledge of sanitary survey process
- Knowledge of watershed protection
- Ability to discriminate between normal and abnormal conditions

Sample Questions for Class I, answers on p. 71

1. A watercourse that flows continuously at all times of the year is called
    a. Intermittent stream
    b. Ephemeral stream
    c. Perennial stream
    d. Natural stream

2. During the night, algae causes the pH of the water to
    a. Increase
    b. Decrease
    c. Remain about the same as during the day
    d. Fluctuate up or down depending on the species of algae that dominate

3. Which of the following contributes to the creation of algae blooms?
    a. Increased turbidity
    b. Increased air temperature
    c. Increased nutrients
    d. Increased wind action

4. Which of the following is found mainly in groundwater sources and forms a precipitate when oxidized?
   a. Hydrogen sulfide
   b. Methane
   c. Radon
   d. Iron

5. The Riparian Doctrine is sometimes called the
   a. Rule of Reasonable Sharing
   b. Appropriation Doctrine
   c. Allocation of Ground Water
   d. Correlative Rights Rule

EVALUATE CHARACTERISTICS OF SOURCE WATER

Sample Questions for Class II, answers on p. 72

1. If a water body has a high salinity and is warm, it will generally be
    a. High in dissolved oxygen
    b. Low in dissolved oxygen
    c. Supersaturated with dissolved oxygen and low in total dissolved solids
    d. Indeterminate for dissolved oxygen and total dissolved solids

2. Which of the following is a major factor that leads to reservoir turnover?
    a. Upper strata becoming warmer and sinking to the bottom
    b. Upper strata becoming cooler and sinking to the bottom
    c. Upper and lower strata reaching the same temperature
    d. Decaying organic matter causing a gaseous movement

3. What does the term *surface runoff* refer to?
    a. Rainwater that soaks into the ground
    b. Rain that returns to the atmosphere from the earth's surface
    c. Surface water that overflows the banks of the rivers during flood stage
    d. Water that flows into the rivers after a rainfall

4. Hikers, bicyclists, all-terrain vehicles, and automobiles can all _____ of watershed land adjacent to a reservoir.
    a. Improve the appearance
    b. Discourage access
    c. Increase the threat of erosion
    d. Limit recreation

5. What is the term for a water-bearing geologic zone composed of material deposited by flowing rivers?
    a. Artesian aquifer
    b. Unconfined aquifer
    c. Consolidated aquifer
    d. Alluvial aquifer

EVALUATE CHARACTERISTICS OF SOURCE WATER

Sample Questions for Class III, answers on p. 73

1. What effect will algae in a reservoir have on dissolved oxygen (DO)?

    a. Lower the DO during the day and increase the DO during the night

    b. Slowly increase the DO during both the day and night

    c. Increase the DO during the day and lower the DO during the night

    d. Slowly decrease the DO during both the day and night

2. What effect does alkalinity have on the use of copper sulfate as an algicide?

    a. Copper sulfate is more effective as alkalinity increases

    b. Copper sulfate is more effective as alkalinity decreases

    c. Alkalinity has no effect

    d. Alkalinity prevents toxicity to fish

3. The amount of water in a water-bearing formation depends on the

    a. Depth of the well

    b. Size of the pump

    c. Porosity of the formation

    d. Type of well casing

4. What is the main characteristic of raw water that enables blue-green algae to grow?

    a. Presence of copper sulfate

    b. Low pH

    c. High hardness

    d. Presence of nutrients

5. A well screen must be installed for which of the following?

    a. All deep wells

    b. Only shallow wells

    c. Consolidated materials

    d. Unconsolidated materials

EVALUATE CHARACTERISTICS OF SOURCE WATER

Sample Questions for Class IV, answers on p. 74

1. What is the recommended loading rate for copper sulfate for algae control at an alkalinity greater than 50 mg/L?

    a. 0.9 lb of copper sulfate per acre of surface area
    b. 1.9 lb of copper sulfate per acre of surface area
    c. 2.4 lb of copper sulfate per acre of surface area
    d. 5.4 lb of copper sulfate per acre of surface area

2. What is the primary origin of coliform bacteria in water supplies?

    a. Natural algae growth
    b. Industrial solvents
    c. Animal or human feces
    d. Acid rain

3. Which of the following best defines the term *specific capacity*?

    a. Amount of water a given volume of saturated rock or sediment will yield to gravity
    b. Amount of water a given volume of saturated rock or sediment will yield to pumping
    c. Rate at which water would flow in an aquifer if the aquifer were an open conduit
    d. Amount of water a well will produce for each foot of drawdown

4. The most common type of well used for public water supply systems is a

    a. Jetted well
    b. Driven well
    c. Drilled well
    d. Bored well

5. Copper sulfate is used for algae control. Your reservoir is 1,200 ft long and 600 ft wide. At an application rate of 5.5 lb per surface acre, how much copper sulfate is required?

    a. 22 lb
    b. 80 lb
    c. 91 lb
    d. 107 lb

MONITOR, EVALUATE, AND ADJUST TREATMENT PROCESSES

	Certification Level			
	I	II	III	IV
Source Treatment				
Algae control		X	X	X
Chemical treatment (copper sulfate)		X	X	X
Stratification control			X	X
Intake structure			X	X
Chemical Treatment/Addition Process				
Chemical addition	X	X	X	X
Fluoridation	X	X	X	X
Chlorine disinfection	X	X	X	X
Chlorine dioxide disinfection	X	X	X	X
Ozone disinfection			X	X
Ultraviolet disinfection	X	X	X	X
pH adjustment	X	X	X	X
Corrosion control	X	X	X	X
Coagulation and Flocculation Process				
Chemical coagulants		X	X	X
Rapid mix units		X	X	X
Flocculation tanks		X	X	X
Coagulation aids			X	X
Clarification/sedimentation process				
Sedimentation basins		X	X	X
Up-flow solids-contact clarification		X	X	X
Inclined-plate sedimentation		X	X	X
Tube sedimentation		X	X	X
Dissolved air flotation			X	X
Filtration Process				
Gravity filtration		X	X	X
Microscreens		X	X	X
Diatomaceous earth filters		X	X	X
Cartridge filters		X	X	X
Slow sand filters		X	X	X
Direct filtration		X	X	X
Pressure or greensand filtration		X	X	X
Filter aids			X	X
Backwash aids			X	X
Other Treatment Processes				
Aeration	X	X	X	X
Packed tower aeration		X	X	X
Ion-exchange/softening		X	X	X
Iron manganese/softening		X	X	X
Lime-soda ash softening		X	X	X
Copper sulfate treatment		X	X	X
Powdered activated carbon		X	X	X
Reverse osmosis		X	X	X

	Certification Level			
	I	II	III	IV
Residuals Disposal				
Discharge to lagoons		X	X	X
Discharge to lagoons and then surface water		X	X	X
Disposal to sanitary sewer		X	X	X
Mechanical dewatering		X	X	X
On-site disposal		X	X	X
Land application		X	X	X
Solids composting		X	X	X

Suggestions for Study:

- Knowledge of chemical properties
- Knowledge of general biology and chemistry
- Knowledge of general electrical and mechanical principles
- Knowledge of normal chemical range
- Knowledge of personal protective equipment
- Knowledge of physical science
- Knowledge of principles of measurement
- Knowledge of proper chemical application, handling, and storage
- Knowledge of proper lifting procedures
- Knowledge of regulations
- Knowledge of water treatment concepts and processes
- Knowledge of water treatment design parameters
- Ability to adjust chemical feed rates
- Ability to adjust flow patterns
- Ability to calculate dosage rates
- Ability to confirm chemical strength
- Ability to diagnose/troubleshoot process units
- Ability to evaluate and adjust process units
- Ability to interpret Material Safety Data Sheets
- Ability to maintain processes in normal operating conditions
- Ability to measure chemical weight/volume
- Ability to perform basic math
- Ability to perform physical measurements
- Ability to perform process control calculations
- Ability to prepare chemicals

MONITOR, EVALUATE, AND ADJUST TREATMENT PROCESSES

Sample Questions for Class I, answers on p. 75

1. Which of the following terms is defined as the killing or inactivation of pathogenic organisms in water?

    a. Sterilization

    b. Pasteurization

    c. Disinfection

    d. Deactivation

2. Free chlorine residual values are based on a contact time of at least

    a. 1 minute

    b. 10 minutes

    c. 30 minutes

    d. 60 minutes

3. What is the typical strength of calcium hypochlorite, i.e., available chlorine range?

    a. 5 to 10%

    b. 45 to 50%

    c. 65 to 70%

    d. 80 to 85%

4. The iron content of a raw water is 1.81 mg/L. What is the percent removal if the finished water contains 0.11 mg/L iron?

    a. 17%

    b. 20%

    c. 60%

    d. 94%

5. Which of the following chemicals will raise pH?

    a. Ferric sulfate

    b. Soda ash

    c. Alum

    d. Carbon dioxide

MONITOR, EVALUATE, AND ADJUST TREATMENT PROCESSES

Sample Questions for Class II, answer on p. 76

1. As water temperature increases, the disinfection action of chlorine will

    a. Increase

    b. Decrease

    c. Double

    d. Indeterminate as it also depends on the pH

2. What effect will caustic soda have on water?

    a. Lower the pH of the water

    b. Buffer the water

    c. Increase the pH of the water

    d. Stabilize the water to a pH of 7

3. In order to prevent freezing, what is the maximum chlorine usage per day from a 150-lb cylinder?

    a. 12 lb

    b. 22 lb

    c. 42 lb

    d. 62 lb

4. Which of the following chemicals is used to soften water through chemical precipitation?

    a. Calcium hydroxide (lime)

    b. Calcium bicarbonate

    c. Calcium chloride

    d. Calcium sulfate

5. How many pounds of 61% calcium hypochlorite are required for a 50-mg/L dosage in a tank that is 110 ft in diameter and has a water level of 19 ft?

    a. 135 lb

    b. 563 lb

    c. 923 lb

    d. 1,124 lb

MONITOR, EVALUATE, AND ADJUST TREATMENT PROCESSES

Sample Questions for Class III, answers on p. 77

1. Mudballs are clumps of
   a. Sand and clay
   b. Filter media and other material
   c. Sand and filter media
   d. Clay and other material

2. Coagulation occurs during the first
   a. 1 to 5 seconds
   b. 10 to 20 seconds
   c. 30 to 60 seconds
   d. 1 to 5 minutes

3. A filter has a surface area of 920 sq ft. What is the filtration rate in gallons per minute per square foot, if it receives a flow of 4,875 gpm?
   a. 2.4 gpm/sq ft
   b. 4.8 gpm/sq ft
   c. 5.3 gpm/sq ft
   d. 9.2 gpm/sq ft

4. A filter should be backwashed when it has a high
   a. Solids loading
   b. Head loss
   c. Chlorine dosage
   d. Influent turbidity

5. Which of the following chemicals decreases corrosion rates?
   a. H_2S
   b. CO_2
   c. $CaCO_3$
   d. O_2

MONITOR, EVALUATE, AND ADJUST TREATMENT PROCESSES

Sample Questions for Class IV, answers on p. 78

1. What it the most important test used to indicate proper sedimentation?
    a. pH
    b. Turbidity
    c. Head loss on filters
    d. Chlorine residual

2. Which of the following is considered an excellent zeta potential?
    a. −1 to −4
    b. −5 to −10
    c. −11 to −20
    d. −21 to −30

3. Which membrane type from the processes listed below has the smallest pore spaces?
    a. Ultrafiltration
    b. Nanofiltration
    c. Reverse osmosis
    d. Microfiltration

4. A treatment plant processes an average of 4,850 gpm. If the lime dosage is 114 g/min, what is the dosage in milligrams per liter?
    a. 0.12 mg/L
    b. 6.22 mg/L
    c. 40.2 mg/L
    d. 51.8 mg/L

5. Find the detention time for the following treatment plant given the following information:
    - 5 flocculation basins each 60 ft by 15 ft, with a water depth of 12 ft
    - 1 sedimentation basin that is 800 ft long, 75 ft wide, with an average water depth of 11 ft
    - 12 filters each 42 ft by 32 ft, with an average water depth of 12 ft
    - Flow is 27.5 mgd
    a. 1.1 hours
    b. 4.9 hours
    c. 5.9 hours
    d. 6.7 hours

LABORATORY ANALYSIS

	Certification Level			
(Collect samples, perform analyses, and interpret results)	I	II	III	IV
Algae analysis			X	X
Alkalinity	X	X	X	X
Aluminum			X	X
Carbon dioxide			X	X
Chlorine demand	X	X	X	X
Chlorine residual	X	X	X	X
Cryptosporidium			X	X
Disinfectant by-products (THM)			X	X
Dissolved oxygen	X	X	X	X
Fluoride concentration	X	X	X	X
Giardia lamblia			X	X
Hardness	X	X	X	X
Inorganic (heavy metal) chemicals	X	X	X	X
Iron/manganese	X	X	X	X
Jar test		X	X	X
Lead/copper	X	X	X	X
Microbiological	X	X	X	X
Nitrate	X	X	X	X
pH	X	X	X	X
Phosphate	X	X	X	X
Radiological parameters	X	X	X	X
Settleable solids			X	X
Synthetic organic chemicals	X	X	X	X
Temperature	X	X	X	X
Turbidity		X	X	X
Volatile organic chemicals	X	X	X	X

Suggestions for Study:

- Knowledge of general biology and chemistry
- Knowledge of laboratory equipment and techniques
- Knowledge of normal characteristics of water
- Knowledge of physical science
- Knowledge of principles of measurement
- Knowledge of proper chemical handling and storage
- Knowledge of proper safety procedures
- Knowledge of proper sampling procedures
- Knowledge of quality control/quality assurance practices
- Knowledge of regulations
- Knowledge of Safe Drinking Water Act
- Knowledge of Standard Methods
- Ability to calibrate instruments
- Ability to follow written procedures
- Ability to interpret Material Safety Data Sheets
- Ability to perform laboratory calculations
- Ability to recognize abnormal analytical results

LABORATORY ANALYSIS

Sample Questions for Class I, answers on p. 80

1. Which of the following should be used by an operator to test for residual chlorine?
    a. DPD (*N, N*-diethyl-*p*-phenylenediamine)
    b. Cresol red
    c. Methyl orange
    d. Sulfuric acid

2. Which one of the following is a major part of a turbidimeter?
    a. Objective nosepiece
    b. Reference electrode
    c. Aspirator
    d. Light source

3. Which of the following parameters is used to indicate the clarity of water?
    a. pH
    b. Chlorine residual
    c. Turbidity
    d. Bacteriological

4. pH is a measure of
    a. Conductivity
    b. Water's ability to neutralize acid
    c. Hydrogen ion concentration
    d. Dissolved solids

5. After chlorination, the free chlorine residual includes
    a. Cl_2, ClO_2, and $HOCl$
    b. OCl^- and $HOCl$
    c. OCl^-, $HOCl$, and Cl_2
    d. ClO_2, $HOCl$, and OCl^-

LABORATORY ANALYSIS

Sample Questions for Class II, answers on p. 81

1. Alkalinity consists predominantly of which of the following?
    a. Bicarbonate, carbonate, and hydroxide
    b. Carbon dioxide and bicarbonate
    c. Carbonate and carbon dioxide
    d. Carbonate, carbon dioxide, and bicarbonate

2. What is the log removal or inactivation required for viruses?
    a. 1.0 log
    b. 2.0 log
    c. 3.0 log
    d. 4.0 log

3. The highest concentration of hypochlorous acid occurs at what pH?
    a. 5
    b. 6
    c. 7
    d. 8

4. Which of the following will most accurately measure 500 mL of solution?
    a. 500-mL beaker
    b. 500-mL Erlenmeyer flask
    c. 500-mL graduated cylinder
    d. 500-mL volumetric flask

5. Which of the following fluoride compounds comes in solution form?
    a. Fluorosilicic acid
    b. Sodium fluoride
    c. Sodium silicofluoride
    d. Fluorine gas

LABORATORY ANALYSIS

Sample Questions for Class III, answers on p. 82

1. Four-log removal is
    a. 90.00%
    b. 99.00%
    c. 99.90%
    d. 99.99%

2. What is the chemical formula for bicarbonate?
    a. H_2CO_3
    b. HCO_3^-
    c. $H_2CO_2^{+2}$
    d. H_2CO

3. The capability of a water or chemical solution to resist a change in pH is called
    a. Chlorine demand
    b. Buffering capacity
    c. Langelier Saturation Index
    d. Zeta potential

4. In coliform analyses using the presence-absence test, a sample should be incubated for
    a. 24 hours at 25°C
    b. 36 hours at 35°C
    c. 24 and 36 hours at 25°C
    d. 24 and 48 hours at 35°C

5. The optimal fluoride concentration in drinking water is set in relation to the
    a. Water alkalinity and pH
    b. Annual average minimum daily water temperature
    c. Annual average maximum daily water temperature
    d. Annual average maximum daily air temperature

LABORATORY ANALYSIS

Sample Questions for Class IV, answers on p. 83

1. The optimum dosage for a polymer can be determined by which of the following tests?

    a. Jar
    b. Marble
    c. Alkalinity
    d. pH

2. In the multiple-tube fermentation method, what is the sequence of growth media?

    a. LTB, EMB, and BGB
    b. LTB, BGB, and EMB
    c. EMB, BGB, and LTB
    d. EMB, LTB, and BGB

3. *Giardia* cysts range in size from

    a. 1 to 2 microns
    b. 2 to 7 microns
    c. 8 to 20 microns
    d. 12 to 20 microns

4. The quantity of oxygen that can remain dissolved in water is related to

    a. Temperature
    b. pH
    c. Turbidity
    d. Alkalinity

5. What is apparent color?

    a. Color in a sample after it is filtered
    b. Color in a sample before it is filtered
    c. Color in a sample after it is disinfected
    d. Color in a sample before it is disinfected

OPERATE EQUIPMENT

	Certification Level			
Operate Equipment	I	II	III	IV
Blowers and compressors	X	X	X	X
Chemical feeders	X	X	X	X
Computers			X	X
Drives	X	X	X	X
Electronic testing equipment	X	X	X	X
Engines	X	X	X	X
Gates	X	X	X	X
Hand tools	X	X	X	X
Hydraulic equipment	X	X	X	X
Instrumentation	X	X	X	X
Motors	X	X	X	X
Off-gas equipment			X	X
Pneumatic equipment			X	X
Power tools	X	X	X	X
Pumps	X	X	X	X
Valves	X	X	X	X

Suggestions for Study:

- Knowledge of function of tools
- Knowledge of general electrical and mechanical principles
- Knowledge of pneumatics
- Knowledge of proper safety procedures
- Knowledge of regulations
- Knowledge of start-up and shut-down procedures
- Knowledge of water treatment concepts
- Ability to adjust equipment
- Ability to evaluate operation of equipment
- Ability to monitor electrical and mechanical equipment

OPERATE EQUIPMENT

Sample Questions for Class I, answers on p. 84

1. After initial full-service operation, grease-lubricated bearings should be regreased at what frequency?

    a. Weekly

    b. Monthly

    c. 3 to 6 months

    d. Annually

2. Check valves are used to prevent

    a. Excessive pump pressure

    b. Priming

    c. Water from flowing in two directions

    d. Water hammer

3. If only two rings of packing are used in the stuffing box, the joints should be

    a. Aligned

    b. Staggered

    c. Large

    d. Made of lead

4. Which of the following is a primary function of couplings?

    a. Compensate for alignment changes

    b. Control motor temperature

    c. Reduce shaft wear

    d. Lubricate motor

5. What is the correct formula for determining watts?

    a. Watts = volts/amps

    b. Watts = (horsepower)(ohms)

    c. Watts = resistance/volts

    d. Watts = (amps)(volts)

OPERATE EQUIPMENT

Sample Questions for Class II, answer on p. 85

1. Centrifugal, positive displacement, and turbine are all
   a. Types of valves
   b. Types of pipe
   c. Water plant chemicals
   d. Types of pumps

2. A hypochlorinator is used to
   a. Measure residual chlorine
   b. Treat iron and turbidity
   c. Feed a liquid chlorine solution into a water supply
   d. Measure an adequate amount of chlorine gas into the supply

3. Water hammer is most likely to be caused by
   a. Dissolved gases in the water
   b. Closing a valve too fast
   c. Tuberculation
   d. Ruptured water line

4. Positive displacement pumps should be operated when
   a. Suction and discharge line valves are closed
   b. Suction and discharge line valves are open
   c. Suction line valves are closed and discharge line valves are open
   d. Suction line valves are open and discharge line valves are closed

5. A 1-ton cylinder filled with chlorine weighs approximately
   a. 2,000 lb
   b. 2,700 lb
   c. 3,700 lb
   d. 4,500 lb

OPERATE EQUIPMENT

Sample Questions for Class III, answers on p. 86

1. Pump seals can be classified as
    a. Balanced or unbalanced
    b. Packing ring or mechanical
    c. Flat and smooth faced or round faced
    d. Factory or custom

2. What is the most probable cause of a pinging sound coming from a pump?
    a. Descaling
    b. Cavitation
    c. Corrosion
    d. Hardness

3. Which of the following best describes the discharge rate of a piston-type pump?
    a. Constant as the speed changes
    b. Precise volume for each stroke
    c. Varies inversely with the head
    d. Varies with the total dynamic head

4. What is the total head in feet for a pump with a total static head of 19 ft and a head loss of 3.7 ft?
    a. 10 ft
    b. 14 ft
    c. 23 ft
    d. 81 ft

5. What is the main purpose of priming?
    a. Ensure the pump operates freely
    b. Compress the air in the cylinder
    c. Replace air with water inside the pump
    d. Wet the packing

OPERATE EQUIPMENT

Sample Questions for Class IV, answers on p. 87

1. What is the purpose of a vacuum regulator?

    a. Relieve excess gas pressure on the chlorinator

    b. Stop flow of chlorine gas if leak develops

    c. Provide a source of air to reduce any excess vacuum on the chlorinator system

    d. Regulate the chlorine feed rate

2. Mechanical seals are more appropriate for pumps operating under which of the following conditions?

    a. High suction head

    b. Low rpm's

    c. High rpm's

    d. High discharge head

3. Which of the following valves is most suitable for a throttling application?

    a. Pressure reducing

    b. Check

    c. Gate

    d. Air-relief

4. The component of a centrifugal pump sometimes installed on the end of the suction pipe in order to hold the priming is the

    a. Casing

    b. Foot valve

    c. Impeller

    d. Lantern ring

5. What is the term for the combined efficiency of a pump and motor that is obtained by multiplying the pump efficiency by the motor efficiency?

    a. Total system efficiency

    b. Well efficiency

    c. Wire-to-water efficiency

    d. Motor-to-pipe efficiency

EVALUATE AND MAINTAIN EQUIPMENT

	Certification Level			
Evaluate operation of equipment:	I	II	III	IV
Check speed of equipment	X	X	X	X
Inspect equipment for abnormal conditions	X	X	X	X
Measure temperature of equipment	X	X	X	X
Read charts	X	X	X	X
Read meters	X	X	X	X
Read pressure gauges	X	X	X	X
Perform preventive maintenance on:				
Chemical feeders	X	X	X	X
Drives	X	X	X	X
Engines	X	X	X	X
Filters	X	X	X	X
Fittings	X	X	X	X
Gates	X	X	X	X
Hydraulic equipment	X	X	X	X
Instrumentation	X	X	X	X
Motors	X	X	X	X
Off-gas equipment			X	X
Pipes	X	X	X	X
Pneumatic equipment			X	X
Pumps	X	X	X	X
Treatment units	X	X	X	X
Valves	X	X	X	X

Suggestions for Study:

- Knowledge of facility operation and maintenance
- Knowledge of general electrical and mechanical principles
- Knowledge of process control instrumentation
- Knowledge of proper safety procedures
- Knowledge of safety regulations
- Knowledge of start-up and shut-down procedures
- Ability to adjust equipment
- Ability to assign work to proper trade
- Ability to calibrate equipment
- Ability to diagnose/troubleshoot process units
- Ability to differentiate between preventative/corrective maintenance
- Ability to discriminate between normal/abnormal equipment conditions
- Ability to follow written procedures
- Ability to monitor electrical and mechanical equipment
- Ability to order necessary spare parts
- Ability to perform general maintenance and repairs
- Ability to record information
- Ability to report findings
- Ability to use hand tools

EVALUATE AND MAINTAIN EQUIPMENT

Sample Questions for Class I, answers on p. 88

1. Opening the discharge valve one full turn on a chlorine cylinder will
   a. Permit maximum discharge
   b. Allow a small flow
   c. Not be enough for efficient operation
   d. Not be safe

2. What of the following best defines the term *"C value"*?
   a. Effectiveness of a disinfectant at a particular concentration as it relates to the destruction of a particular pathogen
   b. Number that describes the smoothness of the interior of a pipe
   c. Corrosiveness of water as calculated by the Langelier index
   d. Factor of the coupon weight after exposure to a water system compared to the coupon's original weight, times 100 to convert to percent

3. Tubercles can form on which of the following types of pipe?
   a. Ductile iron
   b. PVC
   c. Asbestos-cement
   d. Concrete

4. Which of the following is a primary indicator of bearing failure?
   a. Leakage from packing
   b. Reduced pump discharge
   c. Noise from pump
   d. Odor from discharge

5. What is the feet of head at the discharge side of a pump that is pumping against a pressure of 100 psi?
   a. 12.0 ft
   b. 14.6 ft
   c. 43.3 ft
   d. 231 ft

EVALUATE AND MAINTAIN EQUIPMENT

Sample Questions for Class II, answers on p. 89

1. Mechanical seals need replacement when leakage occurs from which of the following?
    a. Pump body
    b. Around the shaft
    c. Volute
    d. Slinger ring

2. Which of the following parts in a centrifugal pump restricts flow between the impeller discharge and suction areas?
    a. Wear rings
    b. Shaft rings
    c. Packing rings
    d. Lantern rings

3. Small gaseous chlorine leaks in and around a chlorinator can be detected by the use of commercial strength
    a. Ammonia
    b. Hypochlorite
    c. Lime
    d. Soda ash

4. Determine the brake horsepower (bhp) if the motor has an efficiency of 91% and the horsepower (hp) is 40.
    a. 30 bhp
    b. 36 bhp
    c. 45 bhp
    d. 67 bhp

5. Packing should be replaced when tightening no longer controls the leakage from the
    a. Mechanical seal
    b. Packing gland
    c. Stuffing box
    d. Shaft sleeve

EVALUATE AND MAINTAIN EQUIPMENT

Sample Questions for Class III, answers on p. 90

1. Air gap separation is measured from the flood level rim of the receiving vessel to the potable water supply pipe. What is the minimum separation of a correctly installed air gap?

    a. 1 in.
    b. 2 in.
    c. 4 in.
    d. 6 in.

2. Which of the following causes air binding?

    a. Lack of liquid in a pump when it is first started
    b. Release of dissolved gases in saturated cold water when pressure decreases in filter beds
    c. Increase of air and thus a decrease of liquid in a pump long after it has been started
    d. Release of dissolved gases in saturated cold water when pressure increases in filter beds

3. Solids-contact units are *most* unstable under which of the following conditions?

    a. Rapid increase in temperature
    b. Rapid increase in pH
    c. Gradual decrease in flow
    d. Gradual decrease in disinfection

4. The least amount of head loss in a pipeline would be the result of a fully open

    a. Angle valve
    b. Check valve
    c. Gate valve
    d. Globe valve

5. During a routine inspection on a centrifugal pump, the operator notices that the bearings are excessively hot. Which of the following is the most likely caused?

    a. Overlubrication
    b. Speed being too slow
    c. Worn impeller
    d. Worn packing

EVALUATE AND MAINTAIN EQUIPMENT

Sample Questions for Class IV, answers on p. 91

1. Electrical demand is
    a. The same as horsepower
    b. Opposition by a circuit to passage of electrons
    c. Amount of power in watts required during a certain time interval
    d. The maximum kilowatt load during a billing period

2. A chlorine evaporator maintains a water bath at
    a. 77–82°F
    b. 110–118°F
    c. 170–180°F
    d. 212–220°F

3. Chlorine ton containers at room temperature will deliver gas at a maximum rate of
    a. 100 pounds per day
    b. 200 pounds per day
    c. 400 pounds per day
    d. 600 pounds per day

4. A water tank has an overflow 115 ft above a nearby fire hydrant. Disregarding friction losses, calculate the pressure at the hydrant when the tank is overflowing.
    a. 25 psi
    b. 50 psi
    c. 133 psi
    d. 356 psi

5. At a pumping station equipped with a centrifugal pump, which of the following can cause the discharge pressure to suddenly increase and the discharge quantity to suddenly decrease?
    a. Pump control valve malfunctioned
    b. Suction valve was closed
    c. Pump amperage was decreased
    d. Voltage was suddenly increased

SAFETY AND EMERGENCY PREPAREDNESS

Safety and Emergency Preparedness	Certification Level			
	I	II	III	IV
Chemical hazard communication	X	X	X	X
Civil disorder/security	X	X	X	X
Confined space entry	X	X	X	X
Electrical grounding	X	X	X	X
Facility upset	X	X	X	X
First aid	X	X	X	X
General safety and health	X	X	X	X
Lifting	X	X	X	X
Lock-out/tag-out	X	X	X	X
Natural disasters	X	X	X	X
Pathogens	X	X	X	X
Personal hygiene	X	X	X	X
Personal protective equipment	X	X	X	X
Respiratory protection	X	X	X	X
Safety meetings	X	X	X	X
Slips, trips, and falls	X	X	X	X
Spill response	X	X	X	X
Transportation	X	X	X	X

Suggestions for Study:

- Knowledge of emergency plan
- Knowledge of potential causes and impact of disasters in facility
- Knowledge of regulations
- Knowledge of risk management
- Ability to assess likelihood of disaster occurring
- Ability to communicate verbally and in writing
- Ability to demonstrate safe work habits
- Ability to generate a written safety program
- Ability to identify potential safety hazards
- Ability to operate and select safety equipment
- Ability to recognize unsafe work conditions

SAFETY AND EMERGENCY PREPAREDNESS

Sample Questions for Class I, answers on p. 92

1. Which of the following diseases is capable of being transmitted by water?
   a. Typhoid ✓
   b. Measles
   c. Encephalitis
   d. Mumps

2. When any piece of electrical equipment is being worked on, the circuit breaker should be
   a. Painted when repair is complete
   b. Videotaped for future reference
   c. De-energized and locked out ✓
   d. Replaced

3. If ammonia vapor is passed over a chlorine leak in a cylinder valve, the presence of the leak is indicated by a
   a. Yellow cloud ✓
   b. White cloud ✓
   c. Gray cloud
   d. Brown cloud

4. Which of the following directly impacts the treatment of drinking water?
   a. Fair Labor Standards Act
   b. Food Security Act
   c. Safe Drinking Water Act ✓
   d. Clean Air Act

5. Corrosive water acting on a customer's plumbing may cause which of the following metals to enter their drinking water?
   a. Lead ✓
   b. Silver
   c. Bismuth
   d. Carbon

SAFETY AND EMERGENCY PREPAREDNESS

Sample Questions for Class II, answers on p. 93

1. Spontaneous combustion can occur when activated carbon is mixed with
    a. Sodium fluoride
    b. Carbon dioxide
    c. Chlorine compounds
    d. Silica sand

2. What is the only acceptable breathing device to wear while handling chlorine leaks?
    a. Activated carbon canister type
    b. Potassium tetroxide canister type
    c. Self-contained breathing apparatus
    d. Oxygen supply apparatus

3. A foul, rotten-egg odor from an aeration treatment unit is an indication that the source water contains
    a. Methane
    b. Carbon dioxide
    c. Hydrogen sulfide
    d. Manganese

4. Of the following types of extinguishers, which should be used on electrical fires?
    a. Water
    b. Soda-acid
    c. Blanket
    d. Carbon dioxide

5. Wear safety goggles when
    a. Handling acid
    b. Doing paperwork
    c. Driving trucks
    d. Measuring turbidity

SAFETY AND EMERGENCY PREPAREDNESS

Sample Questions for Class III, answers on p. 94

1. The proper emergency repair kit for a ton chlorine cylinder is an
    a. Emergency kit A
    b. Emergency kit B
    c. Emergency kit C
    d. Emergency kit T

2. What is the only type of self-contained breathing apparatus that should be used at water plants?
    a. Negative-pressure
    b. Zero-pressure
    c. Positive-pressure
    d. Air-pressure

3. The use of ammonia solution on a chlorine gas leak from the disinfection assembly may cause which of the following?
    a. Explosion
    b. Hydrochloric acid
    c. Toxic gases
    d. Dense white smoke

4. Which of the following substances pose an immediate health threat whenever standards are exceeded?
    a. Benzene and mercury
    b. Coliform and nitrate
    c. Mercury and coliform
    d. Lead and nitrate

5. Ozone contactors must have a system to collect ozone off-gas because ozone is
    a. Toxic
    b. Explosive
    c. Mutagenic
    d. Magnetic

SAFETY AND EMERGENCY PREPAREDNESS

Sample Questions for Class IV, answers on p. 95

1. What is the causative organism for cholera?

    a. *Vibrio*

    b. *Shigella*

    c. *Yersinia*

    d. *Mycobacterium*

2. Which health effect category refers to an organic chemical that is a known carcinogen?

    a. Category I

    b. Category II

    c. Category III

    d. Category IV

3. A room measures 12 ft high, 30 ft long, and 17 ft wide. How many cubic feet per minute of air must a blower in an air exchange unit move to completely change the air every 10 minutes?

    a. 102

    b. 612

    c. 1,020

    d. 6,120

4. One volume of liquid chlorine gas will expand, at room temperature and pressure, to occupy how many volumes of gas?

    a. 16 volumes

    b. 46 volumes

    c. 460 volumes

    d. 960 volumes

5. What is the odor detection limit of chlorine gas?

    a. 0.1 ppm

    b. 0.3 ppm

    c. 0.5 ppm

    d. 1.0 ppm

PERFORM ADMINISTRATIVE DUTIES

	Certification Level			
	I	II	III	IV
Administer safety program	X	X	X	X
Control employee work activities	X	X	X	X
Develop operation and maintenance plan	X	X	X	X
Plan and organize work activities	X	X	X	X
Write reports	X	X	X	X
Respond to complaints	X	X	X	X
Record compliance information	X	X	X	X
Record facility operation information	X	X	X	X
Record financial information	X	X	X	X
Record maintenance information	X	X	X	X
Record personnel information	X	X	X	X

Suggestions for Study:

- Knowledge of facility operation and maintenance
- Knowledge of function of record-keeping system
- Knowledge of monitoring and reporting requirements
- Knowledge of Primary Drinking Water Regulations
- Knowledge of operation and maintenance practices
- Knowledge of principles of general communication
- Knowledge of principles of management
- Knowledge of principles of public relations
- Knowledge of principles of supervision
- Knowledge of record-keeping policies
- Knowledge of regulations
- Ability to communicate verbally and in writing
- Ability to demonstrate safe work habits
- Ability to determine what information needs to be recorded
- Ability to develop a work unit
- Ability to evaluate facility performance
- Ability to follow written procedures
- Ability to generate a written safety program
- Ability to identify potential safety hazards
- Ability to interpret data
- Ability to organize information
- Ability to perform basic math
- Ability to recognize unsafe work conditions
- Ability to review reports
- Ability to transcribe data

PERFORM ADMINISTRATIVE DUTIES

Sample Questions for Class I, answers on p. 96

1. Which of the following is the most common water complaint?
    a. Stained laundry and plumbing fixtures
    b. Objectionable appearance of the water
    c. Objectionable taste and/or odor
    d. Illnesses alleged to be caused by the drinking water

2. According to the Surface Water Treatment Rule, water systems serving a population greater than 500 persons must sample the filtered water for turbidity every
    a. 1 hour
    b. 2 hours
    c. 4 hours
    d. 8 hours

3. For surface water systems without filtration, the Surface Water Treatment Rule requires public water systems to
    a. Provide coagulation and flocculation
    b. Maintain a $C \times T$ value above the minimum value
    c. Continuously sample for total coliforms
    d. Maintain public notification

4. Surface water systems must treat, at a minimum, for
    a. Residual chlorine and *Giardia*
    b. Coliform bacteria and turbidity
    c. Coliform bacteria and residual chlorine
    d. Residual chlorine, turbidity, coliform bacteria, and pH

5. What US agency establishes drinking water standards?
    a. AWWA
    b. USEPA
    c. NIOSH
    d. NSF

PERFORM ADMINISTRATIVE DUTIES

Sample Questions for Class II, answers on p. 97

1. How is a public water system defined?

    a. Must have at least 15 service connections or serve at least 25 individuals daily at least 60 days per year

    b. Must have at least 25 service connections or serve at least 50 individuals daily at least 60 days per year

    c. Must have at least 50 service connections or serve at least 100 individuals daily at least 60 days per year

    d. Must have at least 75 service connections or serve at least 300 individuals daily at least 60 days per year

2. According to the Interim Enhanced Surface Water Treatment Rule, a public water system serving a population of 10,000 or more must maintain the combined effluent turbidity of direct or conventional filtration in 95% of all the measurements taken each month at

    a. ≤0.3 ntu

    b. ≤0.5 ntu

    c. ≤1.0 ntu

    d. ≤5.0 ntu

3. Because it is not possible to routinely test for the presence of specific microorganisms in drinking water, the US Environmental Protection Agency requires, for surface water systems, the use of

    a. Point-of-use treatment

    b. Treatment techniques

    c. Multiple disinfectants

    d. Public notification

4. What does the abbreviation DBP stand for?

    a. Determine by-product

    b. Determinant by-product

    c. Disinfection by-product

    d. Detergent by-product

5. What is the maximum contaminant level goal for chloride?

    a. 2.5 mg/L

    b. 25 mg/L

    c. 250 mg/L

    d. 2,500 mg/L

PERFORM ADMINISTRATIVE DUTIES

Sample Questions for Class III, answers on p. 98

1. What is the maximum contaminant level for nitrate (as nitrogen) in drinking water?
   a. 0.1 mg/L
   b. 0.5 mg/L
   c. 1 mg/L
   d. 10 mg/L

2. Which of the following has both a primary and secondary standard?
   a. Iron
   b. Chromium
   c. Fluoride
   d. Manganese

3. What is the maximum contaminant level goal for known or suspected carcinogens?
   a. Zero
   b. 0.0001 to 0.001
   c. Depends on the chemical
   d. Known or suspected carcinogens only have MCLs

4. Which of the following violations is considered the most serious?
   a. Tier I
   b. Tier II
   c. Tier I-A
   d. Tier II-A

5. An employee receives an hourly wage of $13.25 plus overtime pay of 1.5 times the hourly wage. Overtime pay is given for each hour worked over 40 hours per week. If the employee works 48 hours during a week, what is the compensation before taxes?
   a. $159.00
   b. $530.00
   c. $689.00
   d. $848.00

PERFORM ADMINISTRATIVE DUTIES

Sample Questions for Class IV, answers on p. 99

1. What is the maximum contaminant level goal for lead?
    a. Zero mg/L
    b. 0.015 mg/L
    c. 0.05 mg/L
    d. 0.10 mg/L

2. Records of chemical analyses should be kept for a minimum of
    a. 5 years
    b. 7 years
    c. 10 years
    d. 20 years

3. A water system collects 180 samples for coliform testing. What is the maximum number of samples that may test positive so that the system stays in compliance?
    a. 2
    b. 4
    c. 9
    d. 12

4. The treatment technique required by the US Environmental Protection Agency and used by a surface water system must ensure a log removal for *Giardia* cysts of
    a. 1 log
    b. 2 log
    c. 3 log
    d. 4 log

5. Your department uses 80 units of an item per week. You are required to maintain a 10-week reserve of this item at all times and it requires 4 weeks to obtain a new supply. What is the minimum reorder point?
    a. 320 units
    b. 800 units
    c. 1,120 units
    d. 2,240 units

WATER TREATMENT 39

ADDITIONAL SAMPLE QUESTIONS, answers on p. 100

1. What is the ratio of lime to copper sulfate for controlling algae growth on basin walls?

    a. 1 part lime to 1 part copper sulfate

    b. 1 part lime to 2 parts copper sulfate

    c. 1 part lime to 3 parts copper sulfate

    d. 2 parts lime to 3 parts copper sulfate

2. Copper sulfate is used in surface water reservoirs to control

    a. Emergent weeds

    b. Algae

    c. Mosquito larvae

    d. Snails

3. Which is the correct order (on the average) from smallest to largest for the microorganisms below?

    a. *Giardia* cysts, bacteria, viruses

    b. Viruses, bacteria, *Giardia* cysts

    c. Bacteria, *Giardia* cysts, viruses

    d. *Giardia* cysts, viruses, bacteria

4. What is the usual strength of sodium hypochlorite, i.e., available chlorine?

    a. 5 to 15%

    b. 45 to 50%

    c. 65 to 70%

    d. 80 to 85%

5. A 71.5% calcium hypochlorite solution is used to treat 5.8 mgd. If 237 pounds per day of calcium hypochlorite is used, what is the chlorine dose in milligrams per liter?

    a. 1.7 mg/L

    b. 3.5 mg/L

    c. 5.8 mg/L

    d. 16.9 mg/L

6. A 3.0-ft diameter pipe that is 1,876 ft long was disinfected with chlorine. If 41.33 lb of chlorine were used, what was the dosage in milligrams per liter?

   a. 13 mg/L
   b. 41 mg/L
   c. 50 mg/L
   d. 99 mg/L

7. Trihalomethanes are usually associated with

   a. High levels of algae in a surface water source
   b. Surface water high in organics
   c. Water with organics that has been chlorinated
   d. Groundwater or surface water high in organics

8. Which of the following disinfectants has a long-lasting residual?

   a. Ozone
   b. Ultraviolet light
   c. Chlorine dioxide
   d. Sodium chloride

9. Which of the following substances will reduce the effectiveness of chlorine disinfection?

   a. Turbidity
   b. Color
   c. Radon
   d. Carbon dioxide

10. Which of the following forms of chlorine has no disinfecting capacity?

    a. Hypochlorous acid
    b. Hydrochloric acid
    c. Hypochlorite ion
    d. Dichloramine

11. Find the detention time in minutes for a clarifier that has a diameter of 152 ft and a water depth of 8.22 ft, if the flow rate is 6.8 mgd.

    a. 32 minutes
    b. 236 minutes
    c. 775 minutes
    d. 5,664 minutes

12. A plant is treating water at 21.3 mgd. If lime is being added at a rate of 417.16 g/min, what is the lime dosage in milligrams per liter?

    a. 7.43 mg/L

    b. 8.34 mg/L

    c. 19.6 mg/L

    d. 21.3 mg/L

13. A raw water flow of 17 cfs is prechlorinated with 285 lb of chlorine gas. If the flow is changed to 28 cfs, what should the adjustment be to the chlorinator?

    a. 17 lb of Cl_2

    b. 28 lb of Cl_2

    c. 285 lb of Cl_2

    d. 470 lb of Cl_2

14. An operator mixes 40 lb of lime in a 100-gal tank that contains 80 gal of water. What is the percent of lime in the slurry?

    a. 3%

    b. 6%

    c. 9%

    d. 12%

15. Red water may be caused by iron concentrations above

    a. 0.01 mg/L

    b. 0.03 mg/L

    c. 0.1 mg/L

    d. 0.3 mg/L

16. Which of the following best defines adsorption?

    a. Assimilation of one substance into the body of another by molecular or chemical action

    b. Adhesion of a gas, liquid, or dissolved substance onto the surface or interface zone of another substance

    c. Converting small particles of suspended solids into larger particles by the use of chemicals

    d. Chemical complexing of metallic cations with certain inorganic compounds

17. Hard water scale is usually caused by
    a. Calcium bicarbonate
    b. Calcium carbonate
    c. Magnesium bicarbonate
    d. Magnesium carbonate

18. About how much alkalinity is required for each milligram per liter of alum added to raw water?
    a. 0.5 mg/L
    b. 1.0 mg/L
    c. 1.5 mg/L
    d. 2.0 mg/L

19. A water treatment plant's flocculation-coagulation and sedimentation processes should be checked if which of the following changes?
    a. Turbidity
    b. Chlorine feed rate
    c. Fluoride feed rate
    d. Total trihalomethanes

20. Which of the following is an example of a weighting agent?
    a. Polyelectrolytes
    b. Bentonite clay
    c. Calcium carbonate
    d. Sodium bicarbonate

21. Algae can shorten filter runs by
    a. Clogging the filters
    b. Increasing chlorine demand
    c. Lowering the pH
    d. Increasing turbidity

22. Coagulation is a chemical and physical reaction that converts
    a. Settleable solids into nonsettleable solids
    b. Nonsettleable solids into settleable solids
    c. Dissolved solids into settleable solids
    d. Dissolved solids into a precipitate

23. Water corrosion in metal piping will increase if
    a. The pH is above 7.0 and has low dissolved oxygen levels
    b. The alkalinity is high and water temperature is low
    c. Total dissolved solids are high and the pH is below 7.0
    d. The pH and alkalinity increase

24. What is schmutzdecke?
    a. Settled floc material in a sedimentation basin
    b. Fine sand and a sticky mat of suspended matter that forms on the surface of a sand filter
    c. Adsorption of divalent metal ions onto the surfaces of resins used in ion exchange
    d. Thin layer of protection formed on the inside of metal pipes by the reaction between the alkalinity in the water and zinc orthophosphate

25. What is the cathode?
    a. Negative pole of an electrolytic cell or system
    b. Positive pole of an electrolytic cell or system
    c. Negative pole of a polyelectrolyte
    d. Positive pole of a polyelectrolyte

26. What metals are most likely to leach from household plumbing and cause a health hazard?
    a. Iron and zinc
    b. Iron and copper
    c. Iron and lead
    d. Copper and lead

27. The flow rate over conventional sedimentation basins is commonly designed not to exceed
    a. 12,000 gpd/sq ft
    b. 20,000 gpd/sq ft
    c. 180,000 gpd/sq ft
    d. 250,000 gpd/sq ft

28. If poorly formed floc is leaving the settling basin, which of the following should be done?

    a. Stop the mixer

    b. Increase the coagulant or add a coagulant aid

    c. Increase filter run time

    d. Decrease the chlorination rate

29. Slow sand filters are cleaned by

    a. Air and water backwash

    b. Water backwash alone

    c. Scraping about 1 in. of sand off the top

    d. Surface wash

30. What is the most common filtration rate for slow sand filters?

    a. 0.02 gpm/sq ft

    b. 0.05 gpm/sq ft

    c. 0.1 gpm/sq ft

    d. 0.5 gpm/sq ft

31. Which of the following do long filter runs tend to cause?

    a. Air binding

    b. Slime growths

    c. Mud balls

    d. Media loss

32. A water treatment plant has six filters with an average flow rate of 6.1 gpm/sq ft. If the plant flow is 70 cfs, what is the filtration area of each filter?

    a. 686 sq ft

    b. 861 sq ft

    c. 2,060 sq ft

    d. 5,553 sq ft

33. What type of corrosion in a water system is caused when two different metals come into contact with each other?

    a. Concentration cell corrosion

    b. Uniform corrosion

    c. Galvanic corrosion

    d. Stray current corrosion

34. When should ammonia be added to the water when making disinfectant chloramines to improve disinfection?

    a. Before the chlorine is added
    b. After the chlorine is added
    c. During filtration
    d. Before filtration

35. Alum and ferric sulfate may have poor coagulation due to

    a. High turbidity in the water
    b. Low color in the water
    c. High color in the water
    d. Low alkalinity in the water

36. The process of cathodic protection can be used to reduce or prevent corrosion by protecting metal parts exposed to

    a. Soil
    b. Chloramine
    c. Permafrost
    d. Excessive turbidity

37. Which of the following disinfectants is usually generated on-site?

    a. Chlorine gas
    b. Potassium permanganate
    c. Chlorine dioxide
    d. Sodium hypochlorite

38. If too much potassium permanganate is used to treat water

    a. The sodium level will increase
    b. Disinfection will not be necessary
    c. All contaminants are removed
    d. The water turns pink

39. Which of the following chemicals can be activated for use as a coagulant aid with alum?

    a. Calcium silicate
    b. Sodium silicate
    c. Sodium hypochlorite
    d. Calcium hypochlorite

40. In conventional coagulation, the average time to develop heavy floc particles is?

   a. 1 minute

   b. 10 minutes

   c. 30 minutes

   d. 60 minutes

41. Which of the following is a difference between conventional filters and greensand filters?

   a. Greensand grains are larger than conventional silica grains

   b. Greensand filters remove iron and manganese by adsorption and oxidation

   c. Conventional filters remove iron and manganese by adsorption and oxidation

   d. Conventional beds must be regenerated during backwash

42. Under heavy loading, head loss on a manganese greensand filter quickly becomes excessive because

   a. Greensand grains are smaller than silica sand

   b. Greensand filters are smaller than conventional filters

   c. Greensand filters are not backwashed

   d. Greensand filter runs are longer than conventional filter runs

43. The filter medium in DE filters is

   a. Dielectric earth

   b. Diatomaceous earth

   c. Deionized earth

   d. Disinfected earth

44. Which of the following is a required treatment technique for the control of lead?

   a. Ion exchange

   b. Corrosion control

   c. Lime softening

   d. Activated carbon

45. Bar screens are used to
    a. Remove turbidity
    b. Remove debris
    c. Cover pipes
    d. Size sand

46. Which of the following factors affects the settling rate of particles?
    a. Particle size
    b. Clarifier circumference
    c. Filter surface loading rate
    d. Method of prechlorination

47. Coagulation processes rely on
    a. High-quality water
    b. Higher velocities
    c. Filter valves
    d. The formation of large settleable particles

48. Which of the following substances is the most effective disinfection residual?
    a. Trichloramine
    b. Hypochlorous acid
    c. Chloramine
    d. Hypochlorite ion

49. Water is flowing at a velocity of 1.70 fps in a 10-in. diameter pipe. If the pipe changes from the 10-in. to a 6-in. pipe, what will the velocity be in the 6-in. pipe?
    a. 2.8 fps
    b. 4.6 fps
    c. 10.2 fps
    d. 35.3 fps

50. Which type of solution contains 1 gram equivalent weight of a reactant compound per liter of solution?
    a. Molar solution
    b. Molal solution
    c. Normal solution
    d. Percentage strength solution

51. What term describes a measure of the capacity of water to neutralize strong acids?

    a. Acidity

    b. Alkalinity

    c. Hardness

    d. pH

52. What piece of laboratory glassware is primarily used to mix chemicals and measure approximate volumes?

    a. Beaker

    b. Pipet

    c. Buret

    d. Graduated cylinder

53. What laboratory device sterilizes laboratory apparatus and microbial media by using pressurized steam?

    a. Muffle furnace

    b. Aspirator

    c. Autoclave

    d. Membrane filter

54. Which of the following chemicals cause alkalinity in water?

    a. Calcium carbonate and calcium oxide

    b. Calcium sulfate and sodium sulfate

    c. Magnesium chloride and iron chloride

    d. Sodium chloride and calcium chloride

55. What should the sample volume be when testing for total coliform bacteria?

    a. 100 mL

    b. 250 mL

    c. 500 mL

    d. 1,000 mL

56. Primary drinking water standards require *Giardia* removal at
    a. Two log (99.0%)
    b. Three log (99.9%)
    c. Four log (99.99%)
    d. Five log (99.999%)

57. Recarbonation basins are used to stabilize water after
    a. Filtration
    b. Disinfection
    c. Softening
    d. Coagulation

58. The least reactive metals are called
    a. Anodic metals
    b. Cathodic metals
    c. Galvanic metals
    d. Tempered metals

59. What precipitate is formed in alum coagulation?
    a. Aluminum hydroxide
    b. Complex organo-aluminum compounds
    c. Complex sulfate compounds and aluminum salts
    d. Aluminum salts and organo-sulfate compounds

60. Autoclaving will sterilize
    a. At high temperatures near 600°C
    b. With ultraviolet light
    c. With steam at 121°C and 15 psi
    d. With chlorine gas

61. What is the most common method used to determine if water is close to the equilibrium point?
    a. Marble test
    b. Langelier index
    c. Temperature
    d. Dissolved oxygen

62. Which of the following is used to determine compliance for total coliforms?

    a. Most probable number procedure

    b. Coliforms per 100 mL

    c. Presence-absence method

    d. Heterotrophic plate count

63. What is the percent potassium (K) by weight in potassium permanganate (KMnO$_4$)?

    a. 24.742%

    b. 39.102%

    c. 54.938%

    d. 63.998%

64. Small and medium-size utilities are considered to have optimal corrosion control if they meet the lead and copper action levels for

    a. One sampling period

    b. Two consecutive sampling periods

    c. Three consecutive sampling periods

    d. Four consecutive sampling periods

65. What is pressure head caused by?

    a. Water flow

    b. Water pressure

    c. Water elevation

    d. Gauge pressure

66. What is the motor horsepower (mhp) required if 200 hp is required to move water with a pump with a motor efficiency of 88% and a pump efficiency of 81%? NOTE: The 200 hp in this problem is called the water horsepower (whp). The water horsepower is the actual energy (horsepower) available to pump water.

    a. 166 mhp

    b. 200 mhp

    c. 281 mhp

    d. 355 mhp

67. The flow of electrical current is measured in

    a. Amperes
    b. Ohms
    c. Volts
    d. Watts

68. Unless water cooled, the operating temperature of a mechanical seal in a pump should never exceed

    a. 95°F (35°C)
    b. 120°F (49°C)
    c. 140°F (60°C)
    d. 160°F (71°C)

69. The level in a storage tank (clear well) rises 3.1 ft in 4.5 hours. If the tank has a diameter of 225 ft and the plant is producing 32.4 mgd, what is the average discharge rate of the treated water discharge pumps in gallons per minute?

    a. 3,408 gpm
    b. 15,336 gpm
    c. 19,111 gpm
    d. 22,518 gpm

70. What term describes the condition that exists when the source of the water supply is below the centerline of the pump?

    a. Pressure head
    b. Velocity head
    c. Suction lift
    d. Total discharge head

71. One horsepower is equal to how many foot-pounds per minute?

    a. 746
    b. 3,300
    c. 7,460
    d. 33,000

72. What is the pressure reading at the discharge side of a pump that is pumping against a total head of 100 ft?

   a. 2.3 psi

   b. 4.3 psi

   c. 23.1 psi

   d. 43.3 psi

73. What is the most common use today for a positive-displacement pump?

   a. Raw water intake pump

   b. System booster pump

   c. Chemical feed pump

   d. Filter feed pump

74. What is the purpose of a bypass valve on a larger size gate valve?

   a. Connect a new main to an existing main

   b. Reduce pressure across both sides to ease opening and closing

   c. Increase flow through the main line

   d. Allow easy location of the main valve

75. A corrosion-resistant pipe is used to carry the chlorine gas solution from the chlorine injector to the point of application because the pH range of the solution is

   a. 0–1 standard units

   b. 2–4 standard units

   c. 6–8 standard units

   d. 9–11 standard units

76. Calculate the volume in cubic feet for a pipeline that is 14 in. in diameter and 3,164 ft long.

   a. 1,167 cu ft

   b. 3,164 cu ft

   c. 3,383 cu ft

   d. 6,328 cu ft

77. A well that is 374 ft in depth and 14 in. in diameter requires disinfection. Depth to water from top of casing is 102 ft. If the desired dose is 50 mg/L, how many pounds of calcium hypochlorite (65% available chlorine) are required?

   a. 0.9 lb
   b. 1.4 lb
   c. 10.5 lb
   d. 44.6 lb

78. Water is flowing at a velocity of 3.75 fps in a 10-in. diameter pipe. If the pipe changes from the 10-in. to a 12-in. pipe, what will the velocity be in the 12-in. pipe?

   a. 1.7 fps
   b. 2.6 fps
   c. 3.1 fps
   d. 4.5 fps

79. If a filter is operated so that the pressure in the bed is less than atmospheric, this can lead to short filter runs due to an operating condition known as

   a. Media loss
   b. Mudball formation
   c. Air binding
   d. Gravel displacement

80. What is the head on a system exerting a static pressure of 62 psi?

   a. 27 ft
   b. 89 ft
   c. 143 ft
   d. 175 ft

81. Lead contamination in drinking water will cause children to have

   a. Blue baby syndrome
   b. High blood pressure
   c. Altered mental and physical development
   d. Reduced bone calcium

82. Special care should be taken when using dry alum and quicklime because if mixed together

    a. Explosive hydrogen gas may be released

    b. The mixture will plug effluent weirs

    c. They will create slick areas on the floor

    d. Coagulation will not occur

83. OSHA is the acronym for

    a. Organization for Safe Health Administration

    b. Occupational Safety and Health Administration

    c. Occupation, Safety and Health Act

    d. Organization of State Health Administrators

84. Air scrubbers are

    a. Needed to cleanse sand

    b. For lime room safety

    c. Used to neutralize chlorine leaks

    d. For breathable air

85. The device that changes AC to DC by allowing current flow in only one direction is the

    a. Inverter

    b. Current transformer

    c. Rectifier

    d. Voltage regulator

86. Which of the following is a method of preventing trench cave-in during pipe installation?

    a. Sloping

    b. Scanning

    c. Sequestering

    d. Surging

87. When handling fluoride chemicals, personnel should wear a respirator or mask approved by
    a. OSHA
    b. MSA
    c. EPA
    d. NIOSH

88. What information must be on a warning tag attached to a switch that has been locked out?
    a. Name of person who locked out the switch
    b. Exact time the switch was locked out
    c. Date the switch can be unlocked
    d. Name of shift supervisor

89. Which of the following is colorless, odorless, lighter than air, highly flammable, and sometimes called swamp gas?
    a. Hydrogen sulfide
    b. Methane
    c. Chlorine
    d. Radon

90. Which of the following is colorless, odorless, found mainly in groundwater, and can cause cancer?
    a. Hydrogen sulfide
    b. Methane
    c. Chlorine
    d. Radon

91. What is the major health risk of nitrates?
    a. Methemoglobinemia
    b. Nervous system damage
    c. Prostate cancer
    d. Gastrointestinal effects

92. It is critical to have zero headspace in the sample container when collecting a sample for

    a. Microbiological contaminants
    b. Inorganic chemical contaminants
    c. Organic chemical contaminants
    d. Radiological contaminants

93. According to the Surface Water Treatment Rule requirements, systems serving less than how many people may take grab samples to monitor disinfectant residual?

    a. 3,300
    b. 5,500
    c. 8,500
    d. 10,000

94. A household faucet must remain unused for how many hours before a first draw sample is collected for analyses of lead and copper?

    a. 3 hours
    b. 6 hours
    c. 8 hours
    d. 12 hours

95. Public water systems using surface water as a source must continuously monitor the disinfection residual entering the distribution system if serving more than

    a. 25 people
    b. 300 people
    c. 2,200 people
    d. 3,300 people

96. Under the National Secondary Drinking Water Regulations, what is the secondary maximum contaminant level for total dissolved solids?

    a. 200 mg/L
    b. 250 mg/L
    c. 500 mg/L
    d. 700 mg/L

97. Records for bacteriological analyses should be kept for a minimum of

    a. 5 years

    b. 7 years

    c. 10 years

    d. 20 years

98. Which of the following is **not** an important factor in filter sand selection?

    a. Hardness

    b. Density

    c. Shape

    d. Color

99. What does the abbreviation MCLG stand for?

    a. Minimum concentration level goal

    b. Maximum concentration level goal

    c. Minimum contaminant level goal

    d. Maximum contaminant level goal

100. What is the threshold odor number (TON) that will begin to draw complaints from customers?

    a. 1 TON

    b. 5 TON

    c. 32 TON

    d. 128 TON

MATH FOR MORE PRACTICE, answers on p. 115

1. How many gallons are in 2,167 cu ft?
   a. 260 gal
   b. 295 gal
   c. 16,253 gal
   d. 18,070 gal

2. Convert 8.60 mgd into cubic feet per second.
   a. 8.6 cfs
   b. 13.3 cfs
   c. 798.4 cfs
   d. 19,162.2 cfs

3. Convert 91°F to degrees Celsius.
   a. 33°C
   b. 42°C
   c. 55°C
   d. 74°C

4. If 154 is 72%, what is 100%?
   a. 46
   b. 110
   c. 214
   d. 462

5. What is the average million gallons per day production for a treatment plant given the following data?
   NOTE: All measured values were to the nearest 0.1 mgd.

Mon.	Tues.	Wed.	Thurs.	Fri.	Sat.	Sun.
2.1	2.0	1.8	1.7	1.7	1.6	1.4

   a. 1.7 mgd
   b. 1.8 mgd
   c. 1.9 mgd
   d. 2.0 mgd

6. Calculate the volume in cubic feet for a pipeline that is 14 in. in diameter and 2,156 ft long.

    a. 1,975 cu ft

    b. 2,305 cu ft

    c. 3,172 cu ft

    d. 3,694 cu ft

7. A tank is conical at the bottom and cylindrical at the top. If the diameter of the cylinder is 16 ft with a depth of 30 ft and the cone depth is 10 ft, what is the approximate volume of the tank in gallons?

    a. 20,000 gal

    b. 30,000 gal

    c. 40,000 gal

    d. 50,000 gal

8. A watershed collected 3,267 mil gal for the year. Given the following data, what was the rainfall for a watershed?

    - 11% of the water was collected
    - Watershed is 45 square miles

    a. 25 in.

    b. 38 in.

    c. 45 in.

    d. 67 in.

9. Calculate the detention time in hours for the following water treatment plant:

    - 3 flocculation basins each 48 ft by 20 ft with a water depth of 12.5 ft
    - 1 sedimentation basin that is 565 ft long, 72 ft wide, and has a water depth of 10 ft
    - 8 filters each 40 ft by 28 ft with an average water depth of 10.5 ft and the flow is 18.1 mgd.

    a. 2.4 hours

    b. 5.3 hours

    c. 6.7 hours

    d. 9.4 hours

10. Calculate the pounds per square inch at the bottom of a tank for a water level that is 26.5 ft deep.

    a. 5.4 psi

    b. 11.5 psi

    c. 26.5 psi

    d. 61.2 psi

11. What is the specific gravity for a solution that weighs 10.27 lb/gal?

    a. 1.23

    b. 1.31

    c. 1.39

    d. 1.42

12. What is the flow velocity in feet per second for a 3-in. diameter pipe that delivers 60 gpm?

    a. 2.4 fps

    b. 2.6 fps

    c. 3.3 fps

    d. 3.6 fps

13. If the chlorine dose is 4.25 mg/L and the chlorine residual is 1.20 mg/L, what is the chlorine demand?

    a. 1.20 mg/L

    b. 3.05 mg/L

    c. 4.25 mg/L

    d. 5.45 mg/L

14. What is the approximate number of pounds per day of liquid alum, if the flow rate is 8.2 mgd and the dosage is 12 mg/L? The purity of the alum is 49%.

    a. 17 lb/day

    b. 170 lb/day

    c. 1,700 lb/day

    d. 17,000 lb/day

15. A treatment plant is using 650 lb/day of chlorine gas. If the chlorine demand is 2.3 mg/L and the chlorine residual is 1.2 mg/L, how many million gallons per day are being treated?

    a. 22 mgd
    b. 33 mgd
    c. 65 mgd
    d. 70 mgd

16. A small tank containing 835 gal of water is to be disinfected using a hypochlorite (hypo) solution. If a dosage of 50 mg/L is desired and the available chlorine in the solution is 12%, how much hypochlorite solution should be added in ounces?

    a. 12 oz
    b. 50 oz
    c. 128 oz
    d. 290 oz

17. If a pump discharges 10,350 gal in 3 hours and 45 minutes, how many gallons per minute is the pump discharging?

    a. 43 gpm
    b. 44 gpm
    c. 45 gpm
    d. 46 gpm

18. Find the motor horsepower for a pump station with the following parameters:

    Motor efficiency: 90%   Total Head (TH): 202 ft
    Pump efficiency: 78%   Flow: 1.35 mgd

    a. 47 mhp
    b. 53 mhp
    c. 61 mhp
    d. 68 mhp

19. A channel 7 ft wide has water flowing through it at a depth of 3 ft and a velocity of 2.8 fps. Find the flow through the channel in cubic feet per second.

    a. 49 cfs

    b. 52 cfs

    c. 59 cfs

    d. 65 cfs

20. A rectangular clarifier has a weir length of 100 ft (measured to the nearest foot). What is the weir overflow rate in gallons per day per foot if the flow is 1.7 mgd?

    a. 17 gpd/ft

    b. 170 gpd/ft

    c. 1,700 gpd/ft

    d. 17,000 gpd/ft

21. What is the surface loading rate for a sedimentation basin that is 375 ft by 50 ft (measured to nearest foot), if it is treating an instantaneous flow rate of 17 cfs?

    a. 190 gpd/sq ft

    b. 380 gpd/sq ft

    c. 410 gpd/sq ft

    d. 590 gpd/sq ft

22. Find the drawdown of a well, if the well yields 320 gpm and the specific yield is 18.2.

    a. 17 ft

    b. 18 ft

    c. 19 ft

    d. 20 ft

23. The drawdown worksheet got wet and the operators were unable to read what the pumping water level was. If the static level in the well was 38.23 ft and the drawdown was 19.52 ft, what was the pumping water level in the well?

    a. 18.71 ft

    b. 19.52 ft

    c. 38.23 ft

    d. 57.75 ft

24. A water sample contains 120 mg/L of calcium and 54 mg/L of magnesium. What is the total hardness as CaCO$_3$?

    a. 120 mg/L as CaCO$_3$

    b. 154 mg/L as CaCO$_3$

    c. 174 mg/L as CaCO$_3$

    d. 198 mg/L as CaCO$_3$

25. Water is being pumped from a water source with an elevation of 175 ft to an elevation of 232 ft. What is the total head, if friction and minor head losses are 9 ft?

    a. 57 ft

    b. 66 ft

    c. 78 ft

    d. 83 ft

26. What is the chlorine dosage in milligrams per liter, if 25.7 mgd is treated with 485 lb/day of chlorine?

    a. 2.26 mg/L of chlorine

    b. 2.43 mg/L of chlorine

    c. 2.68 mg/L of chlorine

    d. 2.92 mg/L of chlorine

27. A circular tank has a radius of 15 ft and is 20 ft high. What is the capacity of the tank in cubic feet?

    a. 11,000 cu ft

    b. 12,000 cu ft

    c. 13,000 cu ft

    d. 14,000 cu ft

28. Flow through a channel 4.5 ft wide is 12.8 cfs. If the velocity is 1.9 fps, what is the depth of the water in the channel?

    a. 0.4 ft

    b. 1.5 ft

    c. 1.9 ft

    d. 4.5 ft

29. Find the detention time in hours for a clarifier that has a diameter of 160.0 ft and a water depth of 10.25 ft, if the flow rate is 3.86 mgd.

    a. 4.4 hours
    b. 9.6 hours
    c. 10.3 hours
    d. 14.1 hours

30. A solution was found to be 1.6% alum. What is the parts per million of alum in the solution?

    a. 16 ppm alum
    b. 160 ppm alum
    c. 1,600 ppm alum
    d. 16,000 ppm alum

31. Convert 67 lb/mil gal to milligrams per liter.

    a. 6.7 mg/L
    b. 8.0 mg/L
    c. 9.3 mg/L
    d. 10.6 mg/L

32. Find the capacity of a cylindrical tank in gallons, if it has a diameter of 12.8 ft and has a height of 15.6 ft.

    a. 531 gal
    b. 3,972 gal
    c. 11,061 gal
    d. 15,045 gal

33. Find the detention time in hours for the following treatment plant:
    - 1 sedimentation basin 295 ft long, 80.0 ft wide, and with a water depth of 11.0 ft.
    - 12 filters each 40.0 ft long, 30.0 ft wide, and with an average water depth of 7.00 ft.
    - Flow is 3.50 mgd

    a. 18.5 hours
    b. 19.4 hours
    c. 26.9 hours
    d. 75.4 hours

34. What should the chemical feeder setting be in milliliters per minute for a polymer solution, if the desired dosage is 1.5 mg/L and the treatment plant is treating 14.3 mgd? The specific gravity of the polymer is 1.23.

    a. 10 mL/min
    b. 17 mL/min
    c. 45 mL/min
    d. 180 mL/min

35. A fluoride dose of 1.05 mg/L is desired for treating a flow of 1,750 gpm. How many pounds per day of sodium silicofluoride with a commercial purity of 98% and a fluoride ion content of 60.6% will be required? The water being treated contains 0.15 mg/L fluoride.

    a. 25 lb/day
    b. 32 lb/day
    c. 52 lb/day
    d. 60 lb/day

36. A well is to be disinfected with 62% calcium hypochlorite. The well is 328 ft in depth and is 1.5 ft in diameter. Depth to water from top of casing is 60 ft. If the desired dose is 75 mg/L, how many pounds of calcium hypochlorite are required?

    a. 1.5 lb
    b. 3.6 lb
    c. 4.6 lb
    d. 5.2 lb

37. A filter has an area of 840 sq ft with a backwash pumping rate of 20 cfs. What is the backwash rate in gallons per minute per square foot?

    a. 2 gpm/sq ft
    b. 11 gpm/sq ft
    c. 12 gpm/sq ft
    d. 42 gpm/sq ft

38. Find the approximate amount of manganese removed per year from a plant that treats an average of 5 mgd, if the average manganese concentration is 0.15 ppm and the removal efficiency is 88%?

    a. 2 lb/yr
    b. 20 lb/yr
    c. 200 lb/yr
    d. 2,000 lb/yr

39. Water is flowing at a velocity of 4.5 fps in an 8-in. diameter pipe. If the pipe changes from the 8-in. to a 12-in. pipe, what will the velocity be in the 12-in. pipe?

    a. 1.0 fps

    b. 2.0 fps

    c. 3.0 fps

    d. 4.0 fps

40. A rectangular tank measures 5 ft by 10 ft. Water in the tank is 8.0 ft in depth. What is the pressure in pounds per square inch on the bottom of the tank?

    a. 1.0 psi

    b. 2.3 psi

    c. 3.5 psi

    d. 4.8 psi

41. A conventional treatment plant processes 2,850 gpm on the average for a one-month period. If the lime dosage is 220 grams/min, how many pounds of lime will the plant use in one 30-day month?

    a. 700 lb/month

    b. 1,400 lb/month

    c. 12,300 lb/month

    d. 21,000 lb/month

42. A treatment plant is adding 364 grams/min of soda ash to its treated water. If the plant is producing water at 20 mgd, what is the soda ash usage in pounds per day?

    a. 1,155 lb/day

    b. 3,600 lb/day

    c. 6,343 lb/day

    d. 8,308 lb/day

43. Convert a solution that has 80,350 ppm to percent.

    a. 8%

    b. 80%

    c. 800%

    d. 8,000%

44. Test results of distribution water give a pH of 8.0 and a pH$_s$ of 7.5. What is the Langelier Index and what does this tell you about the distribution system in the area where it was collected?

    a. 0.5: A positive Langelier Index indicates the water is scale forming.

    b. 15.5: A positive Langelier Index indicates the water is corrosive.

    c. –0.5: A negative Langelier Index indicates the water is scale forming.

    d. –15.5: A negative Langelier Index indicates the water is corrosive.

45. Water flowing through a full pipeline has a velocity of 4.6 fps. If the flow through the pipe is 1.6 cfs, what is the diameter in inches of the pipeline?

    a. 4 in.

    b. 6 in.

    c. 8 in.

    d. 10 in.

46. A tank is 40 ft in diameter and is 20 ft tall. If there is 140,900 gal of water in the tank, what is the pounds per square inch at the bottom of the tank?

    a. 2.3 psi

    b. 6.5 psi

    c. 15.1 psi

    d. 18.8 psi

47. A tank is 40 ft in diameter and is 20 ft tall. If there is 140,900 gal of water in the tank, what is the pounds per square inch 5 ft above the bottom of the tank?

    a. 2.2 psi

    b. 4.3 psi

    c. 6.5 psi

    d. 8.7 psi

48. Calculate the theoretical detention time in hours for the following water treatment plant:

    - Flow rate of 11.6 mgd.
    - 5 flocculation basins measuring 45.0 ft by 9.0 ft by 10.0 ft each.
    - 1 sedimentation basin measuring 675 ft by 45.0 ft by 10.0 ft.
    - 8 filters measuring 35.0 ft by 25.0 ft by 12 ft each.
    - Clear well with 2.8 mil gal currently in it.

    a. 10.1 hours

    b. 12.1 hours

    c. 14.3 hours

    d. 16.4 hours

49. A water treatment plant has a filter effluent flow of 7,500 gpm and is being treated with 1,750 gpd of a hypochlorite solution. If the dose is 2.75 mg/L, determine the concentration of the hypochlorite solution in percent.

    a. 0.5%

    b. 0.8%

    c. 1.3%

    d. 1.7%

50. A water district is treating 1,500 mgd of sea water. The total salts contained in the sea water is 2,978 mg/L. How many pounds per year of salts are removed, if 99.3% efficiency is achieved?

    a. 13.5 billion lb/yr

    b. 18.0 billion lb/yr

    c. 26.5 billion lb/yr

    d. 37.0 billion lb/yr

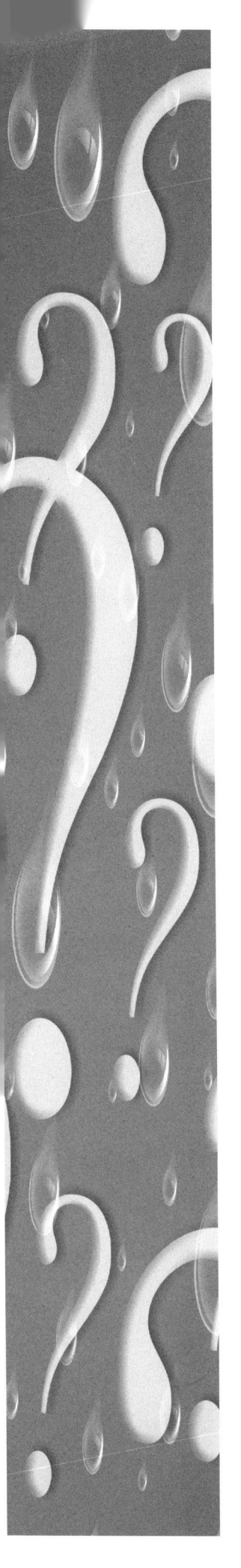

Water Treatment

Answers

EVALUATE CHARACTERISTICS OF SOURCE WATER

Sample Questions for Class I – Answers

1. Answer: **c.** Perennial stream

    Reference: AWWA, *Principles and Practices of Water Supply Operations, Water Sources*, Second Edition, Chapter 3, Page 64.

2. Answer: **b.** Decrease

    Reference: No reference source specified

3. Answer: **c.** Increased nutrients

    Reference: AWWA, *Principles and Practices of Water Supply Operations, Water Sources*, Second Edition, Chapter 3, Page 75.

4. Answer: **d.** Iron

    Reference: AWWA, *Principles and Practices of Water Supply Operations, Water Treatment*, Second Edition, Chapter 9, Page 298.

5. Answer: **a.** Rule of Reasonable Sharing

    Reference: AWWA, *Principles and Practices of Water Supply Operations, Water Sources*, Second Edition, Chapter 5, Page 115.

EVALUATE CHARACTERISTICS OF SOURCE WATER

Sample Questions for Class II – Answers

1. Answer: **b.** Low in dissolved oxygen

   Reference: *AWWA, Principles and Practices of Water Supply Operations, Water Quality,* Second Edition, Chapter 6, Page 154.

2. Answer: **b.** Upper strata becoming cooler and sinking to the bottom

   Reference: *AWWA, Principles and Practices of Water Supply Operations, Water Sources,* Second Edition, Chapter 6, page 124.

3. Answer: **d.** Water that flows into the rivers after a rainfall

   Reference: *AWWA, Principles and Practices of Water Supply Operations, Water Sources,* Second Edition, Chapter 1, Page 3.

4. Answer: **c.** Increase the threat of erosion

   Reference: *AWWA, Principles and Practices of Water Supply Operations, Water Sources,* Second Edition, Chapter 7, Page 174.

5. Answer: **d.** Alluvial aquifer

   Reference: *AWWA, Principles and Practices of Water Supply Operations, Water Sources,* Second Edition, Glossary, Page 177.

EVALUATE CHARACTERISTICS OF SOURCE WATER

Sample Questions for Class III – Answers

1. Answer: **c.** Increase the DO during the day and lower the DO during the night

    Reference: AWWA, *Principles and Practices of Water Supply Operations, Water Sources,* Second Edition, Chapter 5, Page 130.

2. Answer: **b.** Copper sulfate is more effective as alkalinity decreases

    Reference: AWWA, *Principles and Practices of Water Supply Operations, Water Treatment,* Second Edition, Chapter 2, Page 18.

3. Answer: **c.** Porosity of the formation

    Reference: AWWA, *Principles and Practices of Water Supply Operations, Water Sources,* Second Edition, Chapter 1, Page 7.

4. Answer: **d.** Presence of nutrients

    Reference: AWWA, *Principles and Practices of Water Supply Operations, Water Sources,* Second Edition, Chapter 5, Page 137.

5. Answer: **d.** Unconsolidated materials

    Reference: AWWA, *Principles and Practices of Water Supply Operations, Water Sources,* Second Edition, Chapter 2, Page 47.

EVALUATE CHARACTERISTICS OF SOURCE WATER

Sample Questions for Class IV – Answers

1. Answer: **d.** 5.4 lb of copper sulfate per acre of surface area

   Reference: AWWA, *Principles and Practices of Water Supply Operations, Water Treatment,* Second Edition, Chapter 2, Page 18.

2. Answer: **c.** Animal or human feces

   Reference: AWWA, *Principles and Practices of Water Supply Operations, Water Sources,* Second Edition, Chapter 6, Page 143.

3. Answer: **d.** Amount of water a well will produce for each foot of drawdown

   Reference: AWWA, *Principles and Practices of Water Supply Operations, Water Sources,* Second Edition, Chapter 2, Page 27.

4. Answer: **c.** Drilled well

   Reference: AWWA, *Principles and Practices of Water Supply Operations, Water Sources,* Second Edition, Chapter 2, Page 37.

5. Answer: **c.** 91 lb

   Solution: $\dfrac{1200 \text{ ft} \times 600 \text{ ft}}{43{,}560 \text{ sq ft/acre}} \times 5.5 \text{ lb/acre} = 91 \text{ lb}$

   Reference: AWWA, *Principles and Practices of Water Supply Operations, Basic Science Concepts and Applications,* Second Edition, Chapter Mathematics 9, Page 74, and Appendix A, Page 598.

MONITOR, EVALUATE, AND ADJUST TREATMENT PROCESSES

Sample Questions for Class I – Answers

1. Answer: **c.** Disinfection

    Reference: AWWA, *Principles and Practices of Water Supply Operations, Water Treatment,* Second Edition, Chapter 7, Page 161.

2. Answer: **b.** 10 minutes

    Reference: AWWA, *Principles and Practices of Water Supply Operations, Water Treatment,* Second Edition, Chapter 7, Page 219.

3. Answer: **c.** 65 to 70%

    Reference: AWWA, *Principles and Practices of Water Supply Operations, Water Treatment,* Second Edition, Chapter 7, Page 202.

4. Answer: **d.** 94%

    Solution: Equation is: $\% \text{ iron removal} = \frac{(\text{in} - \text{out})}{\text{in}} \times 100\%$

    % iron removal =

    $\frac{(1.81 \text{ mg/L} - 0.11 \text{ mg/L})}{1.81 \text{ mg/L}} \times 100\% = 94\%$ removal efficiency

    Reference: AWWA, *Principles and Practices of Water Supply Operations, Basic Science Concepts and Applications,* Second Edition, Chapter Mathematics 7, Page 65.

5. Answer: **b.** Soda ash. Lime and caustic soda also raise pH.

    Reference: AWWA, *Principles and Practices of Water Supply Operations, Water Treatment,* Second Edition, Chapter 9, Page 275 and 277.

MONITOR, EVALUATE, AND ADJUST TREATMENT PROCESSES

Sample Questions for Class II – Answers

1. Answer: **a.** Increase

   Reference: AWWA, *Principles and Practices of Water Supply Operations, Water Treatment,* Second Edition, Chapter 7, Page 175.

2. Answer: **c.** Increase the pH of the water

   Reference: AWWA, *Principles and Practices of Water Supply Operations, Water Treatment,* Second Edition, Chapter 9, Page 277.

3. Answer: **c.** 42 lb

   Reference: AWWA, *Principles and Practices of Water Supply Operations, Water Treatment,* Second Edition, Chapter 7, Page 212.

4. Answer: **a.** Calcium hydroxide (lime). A number of chemicals can be used to soften water through chemical precipitation. These include, but are not limited to, caustic soda, sodium sulfate, and soda ash.

   Reference: AWWA, *Principles and Practices of Water Supply Operations, Water Treatment,* Second Edition, Chapter 11, Page 318.

5. Answer: **c.** 923 lb

   Solution: First, determine the number of million gallons in the tank with the formula below.

   $$\text{mil gal} = (0.785)(D)^2(\text{depth}) \frac{(7.5 \text{ gal})}{\text{cu ft}} \frac{(1M)}{1{,}000{,}000}$$

   $$\text{mil gal} = (0.785)(110 \text{ ft})(110 \text{ ft})(19 \text{ ft})$$

   $$\frac{(7.5 \text{ gal})}{\text{cu ft}} = \frac{(1M)}{1{,}000{,}000} = 1.35 \text{ mil gal}$$

   Next, use the "pounds/day" equation, but drop the "day" in this problem.

   lb of calcium hypochlorite/day = (mgd) (dosage, mg/L) (8.34 lb/gal)

   lb of calcium hypochlorite = (1.35 mil gal) (50 mg/L) (8.34 lb/gal) = 563 lb

   Since the calcium hypochlorite is not pure (61%), divide the number of pounds by 61% to get the actual number of pounds needed to produce the 50-mg/L dose.

   $$\frac{563 \text{ lb}}{61\%/100\% \text{ Cl}_2 \text{ available}} = 923 \text{ lb of calcium hypochlorite needed}$$

   Reference: AWWA, *Principles and Practices of Water Supply Operations, Basic Science Concepts and Applications,* Second Edition, Chapter Mathematics 10, Page 87, and Chapter Chemistry 7, Page 467 and 471.

MONITOR, EVALUATE, AND ADJUST TREATMENT PROCESSES

Sample Questions for Class III – Answers

1. Answer: **b.** Filter media and other material

   Reference: *AWWA, Principles and Practices of Water Supply Operations, Water Treatment,* Second Edition, Glossary, Page 496.

2. Answer: **a.** 1 to 5 seconds

   Reference: *AWWA, Principles and Practices of Water Supply Operations, Water Treatment,* Second Edition, Chapter 4, Page 76.

3. Answer: **c.** 5.3 gpm/sq ft

   Solution: filtration rate = $\dfrac{\text{flow rate, gpm}}{\text{filter surface area, sq ft}}$

   filtration rate = $\dfrac{4{,}875 \text{ gpm}}{920 \text{ sq ft}}$ = 5.3 gpm/sq ft

   Reference: *AWWA, Principles and Practices of Water Supply Operations, Basic Science Concepts and Applications,* Second Edition, Chapter Mathematics 19, Page 189.

4. Answer: **b.** Head loss. Filters should also be backwashed when effluent turbidity is high.

   Reference: *AWWA, Principles and Practices of Water Supply Operations, Water Treatment,* Second Edition, Chapter 6, Page 139.

5. Answer: **c.** $CaCO_3$

   Reference: *AWWA, Principles and Practices of Water Supply Operations, Water Treatment,* Second Edition, Chapter 9, Page 275.

MONITOR, EVALUATE, AND ADJUST TREATMENT PROCESSES

Sample Questions for Class IV – Answers

1. Answer: **b.** Turbidity

   Reference: AWWA, *Principles and Practices of Water Supply Operations, Water Treatment*, Second Edition, Chapter 5, Page 103.

2. Answer: **a.** −1 to −4

   Reference: AWWA, *Principles and Practices of Water Supply Operations, Water Treatment*, Second Edition, Chapter 4, Page 79.

3. Answer: **c.** Reverse osmosis

   Reference: AWWA, *Principles and Practices of Water Supply Operations, Water Treatment*, Second Edition, Chapter 15, Page 434, and Figure 15-1, Page 433.

4. Answer: **b.** 6.22 mg/L

   Solution: First convert gallons per minute to million gallons per day.
   (4,850 gpm) (1,440 min/day) (1 M/1,000,000) = 6.98 mgd

   Next convert grams per minute of lime to pounds per day.
   (114 g/min) (1 lb/454 g) (1,440 min/day) = 362 lb/day

   To find the dosage, use the "pounds" equation and rearrange to solve for dosage.

   Write the equation: (dosage, lb/day) = (mgd) (dosage, mg/L) (8.34 lb/gal)

   $$\text{dosage, mg/L} = \frac{(\text{dosage, lb/day})}{(\text{mgd})(8.34 \text{ lb/gal})}$$

   $$\text{dosage, mg/L} = \frac{(362 \text{ lb/day})}{(6.98 \text{ mgd})(8.34 \text{ lb/gal})} = 6.22 \text{ mg/L of lime}$$

   Reference: AWWA, *Principles and Practices of Water Supply Operations, Basic Science Concepts and Applications,* Second Edition, Chemistry 7, Page 472.

5. Answer: **c.** 5.9 hours

   Solution:

   First, determine the number of gallons in the five flocculation basins and the sedimentation basin.

The equation is: volume, gal =
(length) (width) (depth) (7.5 gal/cu ft)

volume, gal in floc basins =
(60 ft) (15 ft) (12 ft) (7.5 gal/cu ft) (5 basins) = 405,000 gal

volume, gal in sed basin =
(800 ft) (75 ft) (11 ft) (7.5 gal/cu ft) = 4,950,000 gal

volume, gal in filters =
(42 ft) (32 ft) (12 ft) (7.5 gal/cu ft) (12 filters) = 1,451,520 gal
6,806,520 gal

Next, convert million gallons per day to gallons per hour.

(gph) = (27.5 mgd) (1 day/24 hours) (1,000,000/ 1 M) = 1,145,833 gph

Write the equation with units asked for in the question.

$$\text{detention time, hr} = \frac{\text{volume, gal}}{\text{flow rate, gph}}$$

$$\text{detention time, hr} = \frac{6{,}806{,}520 \text{ gal}}{1{,}145{,}833 \text{ gph}} = 5.9 \text{ hr}$$

Reference: AWWA, *Principles and Practices of Water Supply Operations, Basic Science Concepts and Applications,* Second Edition, Chapter Mathematics 21, Page 200.

LABORATORY ANALYSIS

Sample Questions for Class I – Answers

1. Answer: **a.** DPD

    Reference: *AWWA, Principles and Practices of Water Supply Operations, Water Quality,* Second Edition, Chapter 6, Page 153.

2. Answer: **d.** Light source

    Reference: *AWWA, Principles and Practices of Water Supply Operations, Water Quality,* Second Edition, Chapter 3, Page 98.

3. Answer: **c.** Turbidity

    Reference: *AWWA, Principles and Practices of Water Supply Operations, Water Quality,* Second Edition, Chapter 3, Page 98.

4. Answer: **c.** Hydrogen ion concentration

    Reference: *AWWA, Principles and Practices of Water Supply Operations, Basic Science Concepts and Applications,* Second Edition, Chapter Chemistry 5, Page 426.

5. Answer: **b.** OCl^- and $HOCl$

    Reference: *AWWA, Principles and Practices of Water Supply Operations, Water Treatment,* Second Edition, Chapter 7, Page 172.

WATER TREATMENT 81

LABORATORY ANALYSIS

Sample Questions for Class II – Answers

1. Answer: **a.** Bicarbonate, carbonate, and hydroxide

   Reference: AWWA, *Principles and Practices of Water Supply Operations, Basic Science Concepts and Applications,* Second Edition, Chapter Chemistry 5, Page 427.

2. Answer: **d.** 4.0 log

   Reference: AWWA, *Principles and Practices of Water Supply Operations, Water Treatment,* Second Edition, Chapter 7, Page 209.

3. Answer: **a.** 5

   Reference: AWWA, *Principles and Practices of Water Supply Operations, Water Treatment,* Second Edition, Chapter 7, Page 176, and Figure 7-6.

4. Answer: **d.** 500-mL volumetric flask

   Reference: AWWA, *Principles and Practices of Water Supply Operations, Water Quality,* Second Edition, Chapter 3, Page 66-67.

5. Answer: **a.** Fluorosilicic acid

   Reference: AWWA, *Principles and Practices of Water Supply Operations, Water Treatment,* Second Edition, Chapter 8, Page 238, and Table 8-2.

LABORATORY ANALYSIS

Sample Questions for Class III – Answers

1. Answer: **d.** 99.99%

   Reference: AWWA, *Principles and Practices of Water Supply Operations, Water Treatment,* Second Edition, Chapter 7, Page 209.

2. Answer: **b.** HCO_3^-

   Reference: AWWA, *Principles and Practices of Water Supply Operations, Basic Science Concepts and Applications,* Second Edition, Chapter Chemistry 5, Page 427.

3. Answer: **b.** Buffering capacity

   Reference: AWWA, *Principles and Practices of Water Supply Operations, Water Quality,* Second Edition, Glossary, Page 208.

4. Answer: **d.** 24 and 48 hours at 35°C

   Reference: AWWA, *Principles and Practices of Water Supply Operations, Water Quality,* Second Edition, Chapter 4, Page 110.

5. Answer: **d.** Annual average maximum daily air temperature

   Reference: AWWA, *Principles and Practices of Water Supply Operations, Water Treatment,* Second Edition, Chapter 8, Page 235, and Table 8-1.

LABORATORY ANALYSIS

Sample Questions for Class IV – Answers

1. Answer: **a.** Jar

   Reference: AWWA, *Principles and Practices of Water Supply Operations, Water Treatment,* Second Edition, Chapter 4, Page 77.

2. Answer: **b.** LTB, BGB, and EMB. LTB = Lauryl Tryptose Broth; EMB = Eosin Methylene Blue; BGB = Brilliant Green lactose Bile Broth

   Reference: AWWA, *Principles and Practices of Water Supply Operations, Water Quality,* Second Edition, Chapter 4, Page 108–109, and Figure 4-2.

3. Answer: **c.** 8 to 20 microns

   Reference: AWWA, *Principles and Practices of Water Supply Operations, Water Treatment,* Second Edition, Chapter 15, Page 433, and Figure 15-1.

4. Answer: **a.** Temperature

   Reference: AWWA, *Principles and Practices of Water Supply Operations, Water Quality,* Second Edition, Chapter 6, Page 154.

5. Answer: **b.** Color in a sample before it is filtered

   Reference: AWWA, *Principles and Practices of Water Supply Operations, Water Quality,* Second Edition, Chapter 5, Page 127.

OPERATE EQUIPMENT

Sample Questions for Class I – Answers

1. Answer: **c.** 3 to 6 months

    Reference: AWWA, *Principles and Practices of Water Supply Operations, Water Transmission and Distribution,* Second Edition, Chapter 12, Page 389.

2. Answer: **c.** Water from flowing in two directions

    Reference: AWWA, *Principles and Practices of Water Supply Operations, Water Transmission and Distribution,* Second Edition, Chapter 3, Page 74.

3. Answer: **b.** Staggered

    Reference: AWWA, *Principles and Practices of Water Supply Operations, Water Transmission and Distribution,* Second Edition, Appendix E, Page 565.

4. Answer: **a.** Compensate for alignment changes

    Reference: AWWA, *Principles and Practices of Water Supply Operations, Water Transmission and Distribution,* Second Edition, Appendix E, Page 560.

5. Answer: **d.** Watts = (amps)(volts)

    Reference: AWWA, *Principles and Practices of Water Supply Operations, Basic Science Concepts and Applications,* Second Edition, Chapter Electricity 2, Page 551.

OPERATE EQUIPMENT

Sample Questions for Class II – Answers

1. Answer: **d.** Types of pumps

   Reference: AWWA, *Principles and Practices of Water Supply Operations, Water Transmission and Distribution,* Second Edition, Chapter 12.

2. Answer: **c.** Feed a liquid chlorine solution into a water supply

   Reference: AWWA, *Principles and Practices of Water Supply Operations, Water Treatment,* Second Edition, Chapter 7, Page 203.

3. Answer: **b.** Closing a valve too fast

   Reference: AWWA, *Principles and Practices of Water Supply Operations, Water Transmission and Distribution,* Second Edition, Chapter 12, Page 372.

4. Answer: **b.** Suction and discharge line valves are open

   Reference: AWWA, *Principles and Practices of Water Supply Operations, Water Transmission and Distribution,* Second Edition, Chapter 12, Page 358.

5. Answer: **c.** 3,700 lb

   Reference: AWWA, *Principles and Practices of Water Supply Operations, Water Treatment,* Second Edition, Chapter 7, Page 186.

OPERATE EQUIPMENT

Sample Questions for Class III – Answers

1. Answer: **b.** Packing ring or mechanical

   Reference: AWWA, *Principles and Practices of Water Supply Operations, Water Transmission and Distribution,* Second Edition, Chapter 12, Page 381.

2. Answer: **b.** Cavitation

   Reference: AWWA, *Principles and Practices of Water Supply Operations, Water Transmission and Distribution,* Second Edition, Chapter 12, Page 392.

3. Answer: **b.** Precise volume for each stroke

   Reference: AWWA, *Principles and Practices of Water Supply Operations, Water Transmission and Distribution,* Second Edition, Appendix, Page 594.

4. Answer: **c.** 23 ft

   Solution: total head, ft = total static head, ft + head losses, ft
   total head, ft = 19 ft + 3.7 ft = 22.7 ft, round to 23 ft

   Reference: AWWA, *Principles and Practices of Water Supply Operations, Basic Science Concepts and Applications,* Second Edition, Chapter Hydraulics 4, Page 249.

5. Answer: **c.** Replace air with water inside the pump

   Reference: AWWA, *Principles and Practices of Water Supply Operations, Water Transmission and Distribution,* Second Edition, Chapter 12, Page 386.

OPERATE EQUIPMENT

Sample Questions for Class IV – Answers

1. Answer: **b.** Stop flow of chlorine gas if leak develops

    Reference: AWWA, *Principles and Practices of Water Supply Operations, Water Treatment,* Second Edition, Chapter 7, Page 194 and 195.

2. Answer: **a.** High suction head

    Reference: AWWA, *Principles and Practices of Water Supply Operations, Water Transmission and Distribution,* Second Edition, Chapter 12, Page 382 and 383.

3. Answer: **a.** Pressure reducing

    Reference: AWWA, *Principles and Practices of Water Supply Operations, Water Transmission and Distribution,* Second Edition, Chapter 3, Page 57.

4. Answer: **b.** Foot valve

    Reference: AWWA, *Principles and Practices of Water Supply Operations, Water Transmission and Distribution,* Second Edition, Chapter 12, Page 372.

5. Answer: **c.** Wire-to-water efficiency

    Reference: AWWA, *Principles and Practices of Water Supply Operations, Basic Science Concepts and Applications,* Second Edition, Chapter Hydraulics 6, Page 298.

EVALUATE AND MAINTAIN EQUIPMENT

Sample Questions for Class I – Answers

1. Answer: **a.** Permit maximum discharge

    Reference: AWWA, *Principles and Practices of Water Supply Operations, Water Treatment*, Second Edition, Chapter 7, Page 216.

2. Answer: **b.** Number that describes the smoothness of the interior of a pipe

    Reference: AWWA, *Principles and Practices of Water Supply Operations, Water Transmission and Distribution,* Second Edition, Chapter 1, Page 20.

3. Answer: **a.** Ductile iron

    Reference: AWWA, *Principles and Practices of Water Supply Operations, Water Transmission and Distribution,* Second Edition, Chapter 2, Pages 32–36.

4. Answer: **c.** Noise from pump

    Reference: AWWA, *Principles and Practices of Water Supply Operations, Water Transmission and Distribution,* Second Edition, Chapter 12, Page 384.

5. Answer: **d.** 231 ft

    Solution: (100 psi) (2.31 ft-head/psi) = 231 ft-head

    Reference: AWWA, *Principles and Practices of Water Supply Operations, Basic Science Concepts and Applications,* Second Edition, Chapter Hydraulics 2, Page 227.

EVALUATE AND MAINTAIN EQUIPMENT

Sample Questions for Class II – Answers

1. Answer: **b.** Around the shaft

   Reference: AWWA, *Principles and Practices of Water Supply Operations, Water Transmission and Distribution,* Second Edition, Chapter 12, Page 383-388.

2. Answer: **a.** Wear rings

   Reference: AWWA, *Principles and Practices of Water Supply Operations, Water Transmission and Distribution,* Second Edition, Chapter 12, Page 380.

3. Answer: **a.** Ammonia

   Reference: AWWA, *Principles and Practices of Water Supply Operations, Water Treatment,* Second Edition, Chapter 7, Page 216.

4. Answer: **b.** 36 bhp

   The equation is: bhp = (hp) (% motor efficiency)

   Solution: bhp = (40 hp) (0.91) = 36.4 bhp, round to 36 bhp

   Reference: AWWA, *Principles and Practices of Water Supply Operations, Basic Science Concepts and Applications,* Second Edition, Chapter Hydraulics 6, Page 300.

5. Answer: **b.** Packing gland

   Reference: AWWA, *Principles and Practices of Water Supply Operations, Water Transmission and Distribution,* Second Edition, Chapter 12, Page 387.

EVALUATE AND MAINTAIN EQUIPMENT

Sample Questions for Class III – Answers

1. Answer: **a.** 1 in.

    Reference: AWWA, *Principles and Practices of Water Supply Operations, Water Transmission and Distribution,* Second Edition, Chapter 11, Page 330.

2. Answer: **b.** Release of dissolved gases in saturated cold water when pressure decreases in filter beds

    Reference: AWWA, *Principles and Practices of Water Supply Operations, Water Treatment,* Second Edition, Glossary, Page 485.

3. Answer: **a.** Rapid increase in temperature

    Reference: AWWA, *Principles and Practices of Water Supply Operations, Water Treatment,* Second Edition, Chapter 5, Page 98.

4. Answer: **c.** Gate valve

    Reference: AWWA, *Principles and Practices of Water Supply Operations, Basic Science Concepts and Applications,* Second Edition, Figure H5-2, Page 271.

5. Answer: **a.** Overlubrication

    Reference: AWWA, *Principles and Practices of Water Supply Operations, Water Transmission and Distribution,* Second Edition, Chapter 12, Page 389.

EVALUATE AND MAINTAIN EQUIPMENT

Sample Questions for Class IV – Answers

1. Answer: **c.** Amount of power in watts required during a certain time interval

   Reference: AWWA, *Principles and Practices of Water Supply Operations, Basic Science Concepts and Applications,* Second Edition, Chapter Electricity 2, Page 532.

2. Answer: **c.** 170–180°F

   Reference: AWWA, *Principles and Practices of Water Supply Operations, Water Treatment*, Second Edition, Chapter 7, Page 199.

3. Answer: **c.** 400 pounds per day

   Reference: AWWA, *Principles and Practices of Water Supply Operations, Water Treatment*, Second Edition, Chapter 7, Page 212.

4. Answer: **b.** 50 psi

   Solution: $\dfrac{115 \text{ ft}}{2.31 \text{ ft/psi}} = 50 \text{ psi}$

   Reference: AWWA, *Principles and Practices of Water Supply Operations, Basic Science Concepts and Applications,* Second Edition, Chapter Hydraulics, Page 253.

5. Answer: **a.** Pump control valve malfunctioned

   Reference: AWWA, *Principles and Practices of Water Supply Operations, Water Transmission and Distribution,* Second Edition, Appendix F, Page 573.

SAFETY AND EMERGENCY PREPAREDNESS

Sample Questions for Class I – Answers

1. Answer: **a.** Typhoid

    Reference: AWWA, *Principles and Practices of Water Supply Operations, Water Quality,* Second Edition, Chapter 4, Page 104.

2. Answer: **c.** De-energized and locked out

    Reference: AWWA, *Principles and Practices of Water Supply Operations, Water Transmission and Distribution,* Second Edition, Chapter 13, Page 418.

3. Answer: **b.** White cloud

    Reference: AWWA, *Principles and Practices of Water Supply Operations, Water Treatment,* Second Edition, Chapter 7, Page 216.

4. Answer: **c.** Safe Drinking Water Act

    Reference: AWWA, *Principles and Practices of Water Supply Operations, Water Quality,* Second Edition, Table 1-1, Page 6.

5. Answer: **a.** Lead

    Reference: AWWA, *Principles and Practices of Water Supply Operations, Water Treatment,* Second Edition, Chapter 9, Page 283.

SAFETY AND EMERGENCY PREPAREDNESS

Sample Questions for Class II – Answers

1. Answer: **c.** Chlorine compounds

   Reference: *AWWA, Principles and Practices of Water Supply Operations, Water Treatment,* Second Edition, Chapter 13, Page 397.

2. Answer: **c.** Self-contained breathing apparatus

   Reference: *AWWA, Principles and Practices of Water Supply Operations, Water Treatment,* Second Edition, Chapter 7, Page 225.

3. Answer: **c.** Hydrogen sulfide

   Reference: *AWWA, Principles and Practices of Water Supply Operations, Water Treatment,* Second Edition, Chapter 14, Page 428.

4. Answer: **d.** Carbon dioxide. Dry chemical extinguishers may also be used to extinguish electrical fires.

   Reference: *AWWA, Principles and Practices of Water Supply Operations, Water Transmission and Distribution,* Second Edition, Chapter 13, Page 428.

5. Answer: **a.** Handling acid

   Reference: *No reference available.*

SAFETY AND EMERGENCY PREPAREDNESS

Sample Questions for Class III – Answers

1. Answer: **b.** Emergency kit B

    Reference: AWWA, *Principles and Practices of Water Supply Operations, Water Treatment,* Second Edition, Chapter 7, Page 227.

2. Answer: **c.** Positive-pressure

    Reference: AWWA, *Principles and Practices of Water Supply Operations, Water Treatment,* Second Edition, Chapter 7, Page 225.

3. Answer: **d.** Dense white smoke

    Reference: AWWA, *Principles and Practices of Water Supply Operations, Water Treatment,* Second Edition, Chapter 7, Page 216.

4. Answer: **b.** Coliform and nitrate

    Reference: AWWA, *Principles and Practices of Water Supply Operations, Water Quality,* Second Edition, Chapter 1, Page 16.

5. Answer: **a.** Toxic

    Reference: AWWA, *Principles and Practices of Water Supply Operations, Water Treatment,* Second Edition, Chapter 7, Page 206.

SAFETY AND EMERGENCY PREPAREDNESS

Sample Questions for Class IV – Answers

1. Answer: **a.** *Vibrio*

   Reference: AWWA, *Principles and Practices of Water Supply Operations, Water Treatment*, Second Edition, Chapter 7, Table 7-1, Page 162.

2. Answer: **a.** Category I

   Reference: AWWA, *Principles and Practices of Water Supply Operations, Water Quality,* Second Edition, Chapter 7, Page 171.

3. Answer: **b.** 612

   Solution: 12 ft × 30 ft × 17 ft = 6,120 cu ft

   6,120 cu ft/10 minutes = 612 cfm

   612 cfm provides a change of air every 10 minutes.

   Reference: AWWA, *Principles and Practices of Water Supply Operations, Basic Science Concepts and Applications,* Second Edition, Chapter Math 10, Page 86, and Chapter Hydraulics 6, Page 277.

4. Answer: **c.** 460 volumes

   Reference: AWWA, *Principles and Practices of Water Supply Operations, Water Treatment,* Second Edition, Chapter 7, Page 182.

5. Answer: **b.** 0.3 ppm

   Reference: AWWA, *Principles and Practices of Water Supply Operations, Water Treatment,* Second Edition, Chapter 7, Page 182.

PERFORM ADMINISTRATIVE DUTIES

Sample Questions for Class I – Answers

1. Answer: **c.** Objectionable taste and/or odor

   Reference: *AWWA, Principles and Practices of Water Supply Operations, Water Quality,* Second Edition, Chapter 9, Page 195.

2. Answer: **c.** 4 hours

   Reference: *AWWA, Principles and Practices of Water Supply Operations, Water Treatment,* Second Edition, Chapter 5, Page 140.

3. Answer: **b.** Maintain a $C \times T$ value above the minimum value

   Reference: *AWWA, Principles and Practices of Water Supply Operations, Water Quality,* Second Edition, Chapter 1, Page 22.

4. Answer: **b.** Coliform bacteria and turbidity

   Reference: *AWWA, Principles and Practices of Water Supply Operations, Water Treatment,* Second Edition, Chapter 1, Page 3.

5. Answer: **b.** USEPA

   Reference: *AWWA, Principles and Practices of Water Supply Operations, Water Quality,* Second Edition, Chapter 1, Page 2.

PERFORM ADMINISTRATIVE DUTIES

Sample Questions for Class II – Answers

1. Answer: **a.** Must have at least 15 service connections or serve at least 25 individuals daily at least 60 days per year

    Reference: AWWA, *Principles and Practices of Water Supply Operations, Water Quality*, Second Edition, Chapter 1, Page 4.

2. Answer: **a.** ≤0.3 ntu

    Reference: Reference: Interim Enhanced Surface Water Treatment Rule (IESWTR) 63 FR 69478 69521, December 16, 1998, Vol. 63, No. 241

3. Answer: **b.** Treatment techniques

    Reference: AWWA, *Principles and Practices of Water Supply Operations, Water Quality*, Second Edition, Chapter 1, Page 21.

4. Answer: **c.** Disinfection by-product

    Reference: AWWA, *Principles and Practices of Water Supply Operations, Water Quality*, Second Edition, Chapter 1, Page 28.

5. Answer: **c.** 250 mg/L

    Reference: AWWA, *Principles and Practices of Water Supply Operations, Water Quality*, Second Edition, Chapter 1, Table 1-3, Page 13.

PERFORM ADMINISTRATIVE DUTIES

Sample Questions for Class III – Answers

1. Answer: **d.** 10 mg/L

   Reference: AWWA, *Principles and Practices of Water Supply Operations, Water Quality,* Second Edition, Appendix A, Page 205.

2. Answer: **c.** Fluoride

   Reference: AWWA, *Principles and Practices of Water Supply Operations, Water Quality,* Second Edition, Chapter 1, Page 10, Table 1-3, Page 13, and Appendix A, Page 203.

3. Answer: **a.** Zero

   Reference: AWWA, *Principles and Practices of Water Supply Operations, Water Quality,* Second Edition, Chapter 7, Page 172.

4. Answer: **a.** Tier I

   Reference: AWWA, *Principles and Practices of Water Supply Operations, Water Quality,* Second Edition, Chapter 1, Page 15.

5. Answer: **c.** $689.00

   Solution: (40 hours) ($13.25/hour) + (8 hours) ($13.25/hour) (1.5 overtime pay) = $530 + $159 overtime = $689

   Reference: No reference available

PERFORM ADMINISTRATIVE DUTIES

Sample Questions for Class IV – Answers

1. Answer: **a.** Zero mg/L

    Reference: AWWA, *Principles and Practices of Water Supply Operations, Water Quality,* Second Edition, Appendix A, Page 205.

2. Answer: **c.** 10 years

    Reference: AWWA, *Principles and Practices of Water Supply Operations, Water Quality,* Second Edition, Chapter 1, Table 1-8, Page 21.

3. Answer: **c.** 9

    Solution: (180 samples) (5%) = (180 samples) (0.05) = 9 samples

    Reference: AWWA, *Principles and Practices of Water Supply Operations, Water Quality,* Second Edition, Chapter 4, Page 119.

4. Answer: **c.** 3 log

    Reference: AWWA, *Principles and Practices of Water Supply Operations, Water Quality,* Second Edition, Chapter 1, Page 21.

5. Answer: **c.** 1,120 units

    Solution: (10 weeks) (80 units/week) = 800 units reserve required
    (4 week order time) (80 units/week) = 320 units order time
    800 + 320 = 1,120 units

    Reference: No reference available

ADDITIONAL SAMPLE QUESTIONS – Answers

1. Answer: **a.** 1 part lime to 1 part copper sulfate

   Reference: AWWA, *Principles and Practices of Water Supply Operations, Water Treatment,* Second Edition, Chapter 5, Page 105.

2. Answer: **b.** Algae

   Reference: AWWA, *Principles and Practices of Water Supply Operations, Water Treatment,* Second Edition, Chapter 2, Page 18.

3. Answer: **b.** Viruses, bacteria, *Giardia* cysts

   Reference: AWWA, *Principles and Practices of Water Supply Operations, Water Treatment,* Second Edition, Chapter 15, Figure 15-1, Page 433.

4. Answer: **a.** 5 to 15%

   Reference: AWWA, *Principles and Practices of Water Supply Operations, Water Treatment,* Second Edition, Chapter 7, Page 203.

5. Answer: **b.** 3.5 mg/L

   Solution: First, determine the number of chlorine pounds available.
   available chlorine = (lb used) (% purity) = (237 lb/day) (71.5%/100%) = 169.5 lb/day
   Formula is: (dosage, lb/day) = (mgd) (dosage, mg/L) (8.34 lb/gal)
   Rearrange and solve for dosage.

   $$\text{chlorine dosage, mg/L} = \frac{169.5 \text{ lb/day}}{(5.8 \text{ mgd})(8.34 \text{ lb/gal})} = 3.5 \text{ mg/L}$$

   Reference: AWWA, *Principles and Practices of Water Supply Operations, Basic Science Concepts and Applications,* Second Edition, Chapter Chemistry 7, Page 472.

6. Answer: **c.** 50 mg/L

   Solution: First, find the volume in cubic feet for the pipe.
   Equation is: volume, cu ft = (0.785) (D)^2 (length, ft)
   volume, cu ft = (0.785) (3.0 ft) (3.0 ft) (1,876 ft) = 13,254 cu ft
   Then, determine the number of gallons.
   no. of gal = (13,254 cu ft) (7.5 gal/cu ft) = 99,405 gal, rounded to 99,000
   Convert the number of gallons to million gallons.

   $$\text{mil gal} = \frac{99,000 \text{ gal}}{1,000,000/1 \text{ M}} = 0.099 \text{ mil gal}$$

$$\text{chlorine dosage, mg/L} = \frac{\text{lb of chlorine}}{(\text{mil gal})(8.34 \text{ lb/gal})}$$

chlorine dosage, mg/L =

$$\frac{41.33 \text{ lb}}{(0.099 \text{ mil gal})(8.34 \text{ lb/gal})} = 50.06 \text{ mg/L, round to } 50 \text{ mg/L}$$

Reference: AWWA, *Principles and Practices of Water Supply Operations, Basic Science Concepts and Applications,* Second Edition, Chapter Chemistry 7, Page 473.

7. Answer: **c.** Water with organics that has been chlorinated

 Reference: AWWA, *Principles and Practices of Water Supply and Operations, Water Quality,* Second Edition, Chapter 1, Page 27, and *Water Treatment,* Second Edition. Chapter 7, Page 210.

8. Answer: **c.** Chlorine dioxide

 Reference: AWWA, *Principles and Practices of Water Supply Operations, Water Treatment,* Second Edition, Chapter 7, Table 7-4, Page 167.

9. Answer: **a.** Turbidity. Organic matter, iron, manganese, hydrogen sulfide, and ammonia will also reduce the effectiveness of disinfection.

 Reference: AWWA, *Principles and Practices of Water Supply Operations, Water Treatment,* Second Edition, Chapter 7, Page 177.

10. Answer: **b.** Hydrochloric acid

 Reference: AWWA, *Principles and Practices of Water Supply Operations, Water Treatment,* Second Edition, Chapter 7, Page 171–172.

11. Answer: **b.** 236 minutes

 Solution: First determine the volume in gallons for the clarifier.
 volume, gal = (0.785) (diameter)^2 (depth) (7.48 gal/cu ft)
 volume, gal = (0.785) (152 ft) (152 ft) (8.22 ft) (7.48 gal/cu ft) = 1,115,142 gal
 Then, convert million gallons per day to gallons per minute because detention time is asked for in minutes.
 (6.8 mgd) (1,000,000/1 M) (1 day/1,440 min) = 4,722 gpm

$$\text{Equation is: detention time, min} = \frac{\text{volume, gal}}{\text{flow rate, gpm}}$$

$$\text{detention time, min} = \frac{1{,}115{,}142 \text{ gal}}{4{,}722 \text{ gpm}} = 236 \text{ min}$$

Reference: AWWA, *Principles and Practices of Water Supply Operations, Basic Science Concepts and Applications,* Second Edition, Chapter Mathematics 22, Page 201.

12. Answer: **a.** 7.43 mg/L

    Solution: First, find the pounds per day of lime usage.

    lb/day, lime = (g/min) (1,440 min/day) (1 lb/454 g) = lb/day

    lb/day, lime = (417.16 g/min) (1,440 min/day) (1 lb/454 g) = 1,323 lb/day, round to 1,320 lb/day of lime

    Then, using the "pounds" equation, calculate the dosage in milligrams per
    liter by rearranging the formula and solving for dosage.

    $$\text{dosage, mg/L} = \frac{(\text{dosage, lb/day})}{(\text{mgd})(8.34 \text{ lb/gal})}$$

    $$\text{dosage, mg/L} = \frac{1{,}320 \text{ lb/day}}{(21.3 \text{ mgd})(8.34 \text{ lb/gal})} = 7.43 \text{ mg/L of lime}$$

    Reference: AWWA, *Principles and Practices of Water Supply Operations, Basic Science Concepts and Applications,* Second Edition, Chapter Chemistry 7, Page 467, and Table A-1, Page 609.

13. Answer: **d.** 470 lb of Cl_2

    Solution: Solve by setting up a ratio as follows:

    $$\frac{\text{lb } Cl_2}{\text{flow}_1} = \frac{x \text{ lb } Cl_2}{\text{flow}_2}$$

    Solve for x:

    $$x \text{ lb } Cl_2 = \frac{(285 \text{ lb})(28 \text{ cfs})}{17 \text{ cfs}} = 470 \text{ lb of } Cl_2$$

    Reference: AWWA, *Principles and Practices of Water Supply Operations, Basic Science Concepts and Applications,* Second Edition, Chapter Mathematics 5, Page 53–54.

14. Answer: **b.** 6%

    Solution:

    $$\% \text{ lime} = \frac{(40 \text{ lb})(100\%)}{40 \text{ lb} + (8.34 \text{ lb/gal})(80 \text{ gal})} = \frac{(40 \text{ lb})(100\%)}{40 \text{ lb} + 667.2 \text{ lb}}$$

$$= \frac{(40 \text{ lb})(100\%)}{707.2 \text{ lb}} = 6\% \text{ slurry}$$

Reference: AWWA, *Principles and Practices of Water Supply Operations, Basic Science Concepts and Applications,* Second Edition, Chapter Chemistry 4, Page 399.

15. Answer: **d.** 0.3 mg/L

    Reference: AWWA, *Principles and Practices of Water Supply Operations, Water Quality,* Second Edition, Chapter 6, Page 157.

16. Answer: **b.** Adhesion of a gas, liquid, or dissolved substance onto the surface or interface zone of another substance

    Reference: AWWA, *Principles and Practices of Water Supply Operations, Water Treatment,* Second Edition, Chapter 13, Page 375.

17. Answer: **b.** Calcium carbonate

    Reference: AWWA, *Principles and Practices of Water Supply Operations, Water Treatment,* Second Edition, Chapter 11, Page 316.

18. Answer: **a.** 0.5 mg/L

    Reference: AWWA, *Principles and Practices of Water Supply Operations, Basic Science Concepts and Applications,* Second Edition, Chapter Chemistry 6, Page 437.

19. Answer: **a.** Turbidity

    Reference: AWWA, *Principles and Practices of Water Supply Operations, Water Treatment,* Second Edition, Chapter 5, Page 103.

20. Answer: **b.** Bentonite clay

    Reference: AWWA, *Principles and Practices of Water Supply Operations, Water Treatment,* Second Edition, Chapter 4, Page 59.

21. Answer: **a.** Clogging the filters

    Reference: AWWA, *Principles and Practices of Water Supply Operations, Water Treatment,* Second Edition, Chapter 2, Page 13.

22. Answer: **b.** Nonsettleable solids into settleable solids

    Reference: AWWA, *Principles and Practices of Water Supply Operations, Water Treatment,* Second Edition, Chapter 4, Page 52.

23. Answer: **c.** Total dissolved solids are high and the pH is below 7.0

    Reference: AWWA, *Principles and Practices of Water Supply Operations, Water Treatment,* Second Edition, Chapter 9, Page 268.

24. Answer: **b.** Fine sand and a sticky mat of suspended matter that forms on the surface of a sand filter

Reference: AWWA, *Principles and Practices of Water Supply Operations, Water Treatment,* Second Edition, Chapter 6, Page 118, and Glossary, Page 502.

25. Answer: **a.** Negative pole of an electrolytic cell or system

Reference: AWWA, *Principles and Practices of Water Supply Operations, Water Treatment,* Second Edition, Chapter 9, Page 266, and Glossary, Page 487.

26. Answer: **d.** Copper and lead

Reference: AWWA, *Principles and Practices of Water Supply Operations, Water Treatment,* Second Edition, Chapter 9, Page 264.

27. Answer: **b.** 20,000 gpd/sq ft

Reference: AWWA, *Principles and Practices of Water Supply Operations, Water Treatment,* Second Edition, Chapter 5, Page 101–102.

28. Answer: **b.** Increase the coagulant or add a coagulant aid

Reference: AWWA, *Principles and Practices of Water Supply Operations, Water Treatment,* Second Edition, Chapter 6, Page 103–104.

29. Answer: **c.** Scraping about 1 in. of sand off the top

Reference: AWWA, *Principles and Practices of Water Supply Operations, Water Treatment,* Second Edition, Chapter 6, Page 118.

30. Answer: **b.** 0.05 gpm/sq ft

Reference: AWWA, *Principles and Practices of Water Supply Operations, Water Treatment,* Second Edition, Chapter 6, Page 118.

31. Answer: **b.** Slime growths. Long filter runs also cause a buildup in organic materials and an increase in bacterial populations.

Reference: AWWA, *Principles and Practices of Water Supply Operations, Water Treatment,* Second Edition, Chapter 6, Page 140.

32. Answer: **b.** 861 sq ft

Solution: First, calculate the number of gallons per minute.
(70 cfs) (7.5 gal/cu ft) (60 sec/min) = 31,500 gpm

$$\text{filtration rate} = \frac{\text{flow rate, gpm}}{\text{filter surface area, sq ft}}$$

Rearranging the formula:

$$\text{filter surface area, sq ft} = \frac{(\text{flow rate, gpm})}{(\text{filtration rate})}$$

$$\text{filter surface area, sq ft} = \frac{(31,500 \text{ gpm})}{(6.1 \text{ gpm/sq ft})} = 5,164 \text{ sq ft}$$

filter area for each filter = 5,164 sq ft/6 filters = 861 sq ft for each filter

Reference: AWWA, *Principles and Practices of Water Supply Operations, Basic Science Concepts and Applications,* Second Edition, Chapter Mathematics 19, Page 190.

33. Answer: **c.** Galvanic corrosion

    Reference: AWWA, *Principles and Practices of Water Supply Operations, Water Treatment,* Second Edition, Chapter 9, Page 269.

34. Answer: **b.** After the chlorine is added

    Reference: AWWA, *Principles and Practices of Water Supply Operations, Water Treatment,* Second Edition, Chapter 7, Page 179.

35. Answer: **d.** Low alkalinity in the water

    Reference: AWWA, *Principles and Practices of Water Supply Operations, Water Treatment,* Second Edition, Chapter 4, Page 74.

36. Answer: **a.** Soil

    Reference: AWWA, *Principles and Practices of Water Supply Operations, Water Transmission and Distribution,* Chapter 8, Page 241–245.

37. Answer: **c.** Chlorine dioxide

    Reference: AWWA, *Principles and Practices of Water Supply Operations, Water Treatment,* Second Edition, Chapter 7, Page 208.

38. Answer: **d.** The water turns pink

    Reference: AWWA, *Principles and Practices of Water Supply Operations, Water Treatment,* Second Edition, Chapter 7, Page 207.

39. Answer: **b.** Sodium silicate. Nonionic polyelectrolytes and sodium aluminate can also be used.

    Reference: AWWA, *Principles and Practices of Water Supply Operations, Water Treatment,* Second Edition, Chapter 4, Page 58–59.

40. Answer: **c.** 30 minutes

    Reference: AWWA, *Principles and Practices of Water Supply Operations, Water Treatment,* Second Edition, Chapter 4, Page 76.

41. Answer: **b.** Greensand filters remove iron and manganese by adsorption and oxidation

    Reference: AWWA, *Principles and Practices of Water Supply Operations, Water Treatment,* Second Edition, Chapter 10, Page 303.

42. Answer: **a.** Greensand grains are smaller than silica sand

    Reference: AWWA, *Principles and Practices of Water Supply Operations, Water Treatment,* Second Edition, Chapter 10, Page 303.

43. Answer: **b.** Diatomaceous earth

    Reference: AWWA, *Principles and Practices of Water Supply Operations, Water Treatment,* Second Edition, Chapter 6, Page 152.

44. Answer: **b.** Corrosion control

    Reference: AWWA, *Principles and Practices of Water Supply Operations, Water Quality,* Second Edition, Chapter 1, Page 23.

45. Answer: **b.** Remove debris

    Reference: AWWA, *Principles and Practices of Water Supply Operations, Water Treatment,* Second Edition, Chapter 3, Page 41.

46. Answer: **a.** Particle size. Temperature of the water and wind also affect the settling rate.

    Reference: AWWA, *Principles and Practices of Water Supply Operations, Water Treatment,* Second Edition, Chapter 4, Page 52, Table 4-1.

47. Answer: **d.** The formation of large settleable particles

    Reference: AWWA, *Principles and Practices of Water Supply Operations, Water Treatment,* Second Edition, Chapter 4, Page 54.

48. Answer: **b.** Hypochlorous acid

    Reference: AWWA, *Principles and Practices of Water Supply Operations, Water Treatment,* Second Edition, Chapter 7, Page 172, Table 7-6.

49. Answer: **b.** 4.6 fps

    Solution:
    Flow in 10-in. pipe equals the flow in the 6-in. pipe as the flow must remain constant: $Q_1 = Q_2$

WATER TREATMENT 107

Write the equation as above: (area 1) (velocity 1) = (area 2) (velocity 2)

First, find the diameter for the 6-in. and 10-in. pipes.

diameter for 6-in. = (6-in.) (1 ft/12-in.) = 0.5 ft

diameter for 10-in. = (10-in.) (1 ft/12-in.) = 0.83 ft

Then determine the areas of each size pipe.

area = (0.785) (D)^2

area 1 (6-in.) = (0.785) (0.5 ft) (0.5 ft) = 0.2 sq ft

area 2 (10-in.) = (0.785) (0.83 ft) (0.83 ft) = 0.54 sq ft

Last, substitute areas calculated and known velocity in 10-in. pipe.

(0.2 sq ft) (x, fps) = (0.54 sq ft) (1.70 fps)

Solve for x:

$$x, \text{fps} = \frac{(0.54 \text{ sq ft})(1.70 \text{ fps})}{(0.2 \text{ sq ft})} = 4.6 \text{ fps in the 6-in. pipe}$$

Reference: AWWA, *Principles and Practices of Water Supply Operations, Basic Science Concepts and Applications,* Second Edition, Hydraulics 7, Page 320, and Mathematics 5, Page 54.

50. Answer: **c.** Normal solution

    Reference: AWWA, *Principles and Practices of Water Supply Operations, Basic Science Concepts and Applications,* Second Edition, Chemistry 4, Page 411.

51. Answer: **b.** Alkalinity

    Reference: AWWA, *Principles and Practices of Water Supply Operations, Basic Science Concepts and Applications,* Second Edition, Chemistry 5, Page 427.

52. Answer: **a.** Beaker

    Reference: AWWA, *Principles and Practices of Water Supply Operations, Water Quality,* Second Edition, Chapter 3.

53. Answer: **c.** Autoclave

    Reference: AWWA, *Principles and Practices of Water Supply Operations, Water Quality,* Second Edition, Chapter 3, Page 79.

54. Answer: **a.** Calcium carbonate and calcium oxide

    Reference: AWWA, *Principles and Practices of Water Supply Operations, Basic Science Concepts and Applications,* Second Edition, Chemistry 5, Page 427.

55. Answer: **a.** 100 mL

    Reference: AWWA, *Principles and Practices of Water Supply Operations, Water Quality,* Second Edition, Chapter 4, Page 115.

56. Answer: **b.** Three log (99.9%)

    Reference: *AWWA, Principles and Practices of Water Supply Operations, Water Treatment,* Second Edition, Chapter 7, Page 209.

57. Answer: **c.** Softening

    Reference: *AWWA, Principles and Practices of Water Supply Operations, Water Treatment,* Second Edition, Chapter 11, Page 343.

58. Answer: **b.** Cathodic metals

    Reference: *AWWA, Principles and Practices of Water Supply Operations, Water Treatment,* Second Edition, Chapter 9, Table 9-2, Page 269.

59. Answer: **a.** Aluminum hydroxide

    Reference: *AWWA, Principles and Practices of Water Supply Operations, Water Treatment,* Second Edition, Chapter 4, Page 56.

60. Answer: **c.** Sterilize with steam at 121°C and 15 psi

    Reference: *AWWA, Principles and Practices of Water Supply Operations, Water Quality,* Second Edition, Chapter 3, Page 79.

61. Answer: **b.** Langelier index

    Reference: *AWWA, Principles and Practices of Water Supply Operations, Water Quality,* Second Edition, Chapter 6, Page 146.

62. Answer: **c.** Presence-absence method

    Reference: *AWWA, Principles and Practices of Water Supply Operations, Water Quality,* Second Edition, Chapter 4, Page 108–112.

63. Answer: **a.** 24.742%

    Solution: The equation for calculating the % K in potassium permanganate is:

    $$\% \text{ K} = \frac{\text{molecular wt of K (100\%)}}{\text{molecular wt of KMnO}_4}$$

    First, determine the molecular weight of each of the elements in the compound.

Element	Number of atoms		Atomic Wt		Molecular Wt
K	1	×	39.102	=	39.102
Mn	1	×	54.938	=	54.938
O	4	×	15.9994	=	63.9976
			molecular wt of KMnO$_4$	=	158.0376

Substituting: %K = $\dfrac{39.102\ (100\%)}{158.0376}$ = 24.742 %

Reference: AWWA, *Principles and Practices of Water Supply Operations, Basic Science Concepts and Applications,* Second Edition, Chemistry 3, Page 382, and Appendix B, Table B-1, Page 622.

64. Answer: **b.** Two consecutive sampling periods

    Reference: AWWA, *Principles and Practices of Water Supply Operations, Water Quality,* Second Edition, Chapter 1, Page 25.

65. Answer: **c.** Water elevation

    Reference: AWWA, *Principles and Practices of Water Supply Operations, Basic Science Concepts and Applications,* Second Edition, Hydraulics 4, Page 246.

66. Answer: **c.** 281 mhp

    Solution: The equation is: motor hp = $\dfrac{(whp)}{(motor\ effic.)(pump\ effic.)}$

    motor hp =

    $\dfrac{(200\ whp)}{(88\%\ motor\ effic.)(81\%\ pump\ effic.)} = \dfrac{(200\ whp)}{(0.88\ motor\ effic.)(0.81\ pump\ effic.)}$

    motor hp = 281 mhp

    Reference: AWWA, *Principles and Practices of Water Supply Operations, Basic Science Concepts and Applications,* Second Edition, Hydraulics 6, Example 20, Page 301–302.

67. Answer: **a.** Amperes

    Reference: AWWA, *Principles and Practices of Water Supply Operations, Basic Science Concepts and Applications,* Second Edition, Electricity 1, Page 526.

68. Answer: **d.** 160°F (71°C)

    Reference: AWWA, *Principles and Practices of Water Supply Operations, Water Transmission and Distribution,* Second Edition, Chapter 12, Page 388.

69. Answer: **c.** 19,111 gpm

    Solution: First, find the water production during the 4.5-hr interval.
    water produced in 4.5-hr interval = (32.4 mgd) (1,000,000 gal/M) (1 day/24 hr) (4.5 hr) = 6,075,000 gal, round to 6,080,000 gal

    Next, find the number of gallons contained in the 3.1-ft rise in water level.

volume, tank = (0.785) (D)² (height)

volume of 3.1 ft in 225 ft diameter tank = (0.785) (225 ft) (225 ft)
(3.1 ft) (7.5 gal/cu ft) = 923,969 gal, round to 920,000 gal

Production minus the rise in level is the amount the discharge pumps had to send into the distribution system.

total gal discharge pumps moved in 4.5 hr = 6,080,000 gal – 920,000 gal
= 5,160,000 gal

Next, find the number of minutes in 4.5 hr.
(4.5 hr) (60 min/hr) = 270 min

Last, divide the number of gallons the discharge pumps moved by the time in minutes.

discharge pumps, gpm = 5,160,000 gal/270 min = 19,111 gpm

Reference: AWWA, *Principles and Practices of Water Supply Operations, Basic Science Concepts and Applications,* Second Edition, Mathematics 10, Page 85, and Hydraulics 6, Page 273.

70. Answer: **c.** Suction lift

Reference: AWWA, *Principles and Practices of Water Supply Operations, Basic Science Concepts and Applications,* Second Edition, Hydraulics 6, Figure H6-8B, Page 282.

71. Answer: **d.** 33,000

Reference: AWWA, *Principles and Practices of Water Supply Operations, Basic Science Concepts and Applications,* Second Edition, Hydraulics 6, Page 292, and Table A-1, Page 604.

72. Answer: **d.** 43.3 psi

Solution: $\dfrac{100 \text{ ft-head}}{2.31 \text{ psi/ft-head}} = 43.3 \text{ psi}$

Reference: AWWA, *Principles and Practices of Water Supply Operations, Basic Science Concepts and Applications,* Second Edition, Hydraulics 2, Page 227.

73. Answer: **c.** Chemical feed pump

Reference: AWWA, *Principles and Practices of Water Supply Operations, Water Transmission and Distribution,* Second Edition, Chapter 12, Page 369.

74. Answer: **b.** Reduce pressure across both sides to ease opening and closing

Reference: AWWA, *Principles and Practices of Water Supply Operations, Water Transmission and Distribution,* Second Edition, Chapter 3, Page 65.

WATER TREATMENT   111

75. Answer: **b.** 2–4 standard units

   Reference: AWWA, *Principles and Practices of Water Supply Operations, Water Treatment,* Second Edition, Chapter 7, Page 195.

76. Answer: **c.** 3,383 cu ft

   Solution: First, convert the diameter to feet.

   $(14 \text{ in.}) \dfrac{(1 \text{ ft})}{12 \text{ in.}} = 1.167$ ft (diameter)

   Equation: volume, cu ft = (0.785) (D)^2 (length)
   volume, cu ft = (0.785) (1.167 ft) (1.167 ft) (3,164 ft)
   volume = 3,383 cu ft

   Reference: AWWA, *Principles and Practices of Water Supply Operations, Basic Science Concepts and Applications,* Second Edition, Mathematics 10, Page 87.

77. Answer: **b.** 1.4 lb

   Solution: First, find the length (in feet) of water in the casing.
   length of water-filled casing = depth of well − depth of water to top of casing
   length of water-filled casing = 374 ft − 102 ft = 272 ft
   Then, convert the diameter from inches to feet.

   diameter, ft = $\dfrac{14 \text{ in.}}{12 \text{ in./ft}} = 1.167$ ft

   Next, determine the volume in gallons of water in the well casing using the following formula:
   volume, in gal = (0.785) (D)^2 (length) (7.5 gal/cu ft)
   volume, in gal = (0.785) (1.167 ft) (1.167 ft) (272 ft) (7.5 gal/cu ft) = 2,181 gal

   Next, determine the number of million gallons.
   mi/gal = (2,181 gal) (1 M/1,000,000) = 0.002181 mil gal

   Last, using the "pounds" formula, calculate the number of pounds of calcium hypochlorite.

   lb of calcium hypochlorite =

   $\dfrac{(0.002181 \text{ mil gal})(50 \text{ mg/L})(8.34 \text{ lb/gal})}{(65\%/100\% \text{ available chlorine})} = 1.4 \text{ lb}$

   Reference: AWWA, *Principles and Practices of Water Supply Operations, Basic Science Concepts and Applications,* Second Edition, Mathematics 10, Page 87, and Chemistry 7, Page 472–473.

112 CERTIFICATION STUDY GUIDE

78. Answer: **b.** 2.6 fps

    Solution: Flow in 10-in. pipe equals flow in the 12-in. pipe as the flow must remain constant: $Q_1 = Q_2$.

    Write the equation: (area 1) (velocity 1) = (area 2) (velocity 2)

    First, find the diameter for the 12-in. and 10-in. pipes in feet.

    diameter for 12-in. = (12-in.) (1 ft/12-in.) = 1.0 ft

    Diameter for 10-in. = (10-in.) (1 ft/12-in.) = 0.83 ft

    Then determine the areas of each size pipe.

    area = (0.785) (D)^2

    area 1 (12-in.) = (0.785) (1.0 ft) (1.0 ft) = 0.785 sq ft

    area 2 (10-in.) = (0.785) (0.83 ft) (0.83 ft) = 0.54 sq ft

    Last, substitute areas calculated and known velocity in 10-in. pipe.

    (0.785 sq ft) (x, fps) = (0.54 sq ft) (3.75 fps)

    Solve for x.

    $$x, \text{fps} = \frac{(0.54 \text{ sq ft})(3.75 \text{ fps})}{(0.785 \text{ sq ft})} = 2.6 \text{ fps in 12-in. pipe}$$

    Reference: AWWA, *Principles and Practices of Water Supply Operations, Basic Science Concepts and Applications,* Second Edition, Hydraulics 7, Example 9, Page 328-329.

79. Answer: **c.** Air binding

    Reference: AWWA, *Principles and Practices of Water Supply Operations, Water Treatment,* Second Edition, Chapter 6, Page 150.

80. Answer: **c.** 143 ft

    Solution: 1 psig = 2.31 ft head
    (62 psig) (2.31 ft/psig) = 143 ft

    Reference: AWWA, *Principles and Practices of Water Supply Operations, Basic Science Concepts and Applications,* Second Edition, Hydraulics 2, Page 227, Example 2.

81. Answer: **c.** Altered mental and physical development

    Reference: AWWA, *Principles and Practices of Water Supply Operations, Water Quality,* Second Edition, Chapter 1, Page 24.

82. Answer: **a.** Explosive hydrogen gas may be released

    Reference: AWWA, *Principles and Practices of Water Supply Operations, Water Treatment,* Second Edition, Chapter 4, Page 81.

WATER TREATMENT 113

83. Answer: **b.** Occupational Safety and Health Administration

 Reference: AWWA, *Principles and Practices of Water Supply Operations, Water Treatment,* Second Edition, Chapter 7, Page 206, and *Water Transmission and Distribution,* Second Edition, Chapter 11, Page 348.

84. Answer: **c.** Used to neutralize chlorine leaks

 Reference: AWWA, *Principles and Practices of Water Supply Operations, Water Treatment,* Second Edition, Chapter 7.

85. Answer: **c.** Rectifier

 Reference: AWWA, *Principles and Practices of Water Supply Operations, Basic Science Concepts and Applications,* Second Edition, Electricity 3, Page 563.

86. Answer: **a.** Sloping

 Reference: AWWA, *Principles and Practices of Water Supply Operations, Water Transmission and Distribution,* Second Edition, Chapter 4, Page 98.

87. Answer: **d.** NIOSH (National Institute for Occupational Safety and Health)

 Reference: AWWA, *Principles and Practices of Water Supply Operations, Water Treatment,* Second Edition, Chapter 8, Page 260.

88. Answer: **a.** Name of person who locked out the switch

 Reference: AWWA, *Principles and Practices of Water Supply Operations, Water Transmission and Distribution,* Second Edition, Chapter 12, Page 394.

89. Answer: **b.** Methane

 Reference: AWWA, *Principles and Practices of Water Supply Operations, Water Treatment,* Second Edition, Glossary, Page 496.

90. Answer: **d.** Radon

 Reference: AWWA, *Principles and Practices of Water Supply Operations, Water Quality,* Second Edition, Chapter 8, Page 187–188.

91. Answer: **a.** Methemoglobinemia (also called blue baby syndrome)

 Reference: AWWA, *Principles and Practices of Water Supply Operations, Water Quality,* Second Edition, Appendix A, Page 205.

92. Answer: **c.** Organic chemical contaminants

    Reference: *AWWA, Principles and Practices of Water Supply Operations, Water Quality,* Second Edition, Chapter 7, Page 176.

93. Answer: **a.** 3,300

    Reference: *AWWA, Principles and Practices of Water Supply Operations, Water Treatment,* Second Edition, Chapter 7, Page 210.

94. Answer: **b.** 6 hours

    Reference: *AWWA, Principles and Practices of Water Supply Operations, Water Quality,* Second Edition, Chapter 1, Page 25.

95. Answer: **d.** 3,300 people

    Reference: *AWWA, Principles and Practices of Water Supply Operations, Water Quality,* Second Edition, Chapter 1, Page 22, and *Water Treatment,* Second Edition, Chapter 7, Page 210.

96. Answer: **c.** 500 mg/L

    Reference: *AWWA, Principles and Practices of Water Supply Operations, Water Quality,* Second Edition, Chapter 1, Table 1-3, Page 13.

97. Answer: **a.** 5 years

    Reference: *AWWA, Principles and Practices of Water Supply Operations, Water Quality,* Second Edition, Chapter 1, Page 21, Table 1-8.

98. Answer: **d.** Color

    Reference: *AWWA, Principles and Practices of Water Supply Operations, Water Treatment,* Second Edition, Chapter 6, Page 123.

99. Answer: **d.** Maximum contaminant level goal

    Reference: *AWWA, Principles and Practices of Water Supply Operations, Water Quality,* Second Edition, Chapter 1, Page 8-9.

100. Answer: **b.** 5 TON

    Reference: *AWWA, Principles and Practices of Water Supply Operations, Water Quality,* Second Edition, Chapter 5, Page 134.

WATER TREATMENT 115

MATH FOR MORE PRACTICE – Answers

1. Answer: **c.** 16,253 gal

    Solution: $\dfrac{(7.5 \text{ gal})}{1 \text{ cu ft}}$ (2,167 cu ft) = 16,252.5 gal, round to 16,253 gal

2. Answer: **b.** 13.3 cfs

    Solution:

    $$\dfrac{(8.60 \text{ mgd})}{} \dfrac{(1{,}000{,}000 \text{ gal})}{1 \text{ mil gal}} \dfrac{(1 \text{ cu ft})}{7.5 \text{ gal}} \dfrac{(1 \text{ day})}{1{,}440 \text{ min}} \dfrac{(1 \text{ min})}{60 \text{ sec}} = 13.3 \text{ cfs}$$

3. Answer: **a.** 33°C

    Solution: The equation for Celsius is: °C = 5/9 (°F – 32 °F)
    °C = 5/9 (91°F – 32°F) = 5/9 (59°F) = 33°C

4. Answer: **c.** 214

    Solution: Writing a ratio can solve this problem.
    154/72% = x/100%

    $x = \dfrac{(100\%)(154)}{72\%}$ = 213.9, round to 214

    An easier way to solve the problem is to know that the number for 100% must be larger than 154. If we divide by the decimal for 72%, we get the same answer, and it is a little faster to solve.
    x = 154/0.72 = 214

5. Answer: **b.** 1.8 mgd

    Solution: average mgd produced = $\dfrac{\text{sum of mgd used each day}}{\text{total time, days}}$

    avg. mgd produced = $\dfrac{2.1 + 2.0 + 1.8 + 1.7 + 1.7 + 1.6 + 1.4}{7 \text{ days}}$ = 1.76 mgd, round to 1.8 mgd

6. Answer: **b.** 2,305 cu ft

    Solution: First, convert the diameter to feet.

    (14 in.) $\dfrac{(1 \text{ ft})}{12 \text{ in.}}$ = 1.167 ft (diameter)

    Formula for the volume of a pipe in cubic feet is: 0.785 (D)^2 (length):

volume = (0.785) (1.167 ft) (1.167 ft) (2,156 ft)
volume = 2,305 cu ft

7. Answer: **d.** 50,000 gal

    Solution: First, find the volume of the cone in cubic feet
    volume, cu ft = ⅓πr² (depth): Where the radius = diameter/2 = 16 ft/2 = 8 ft
    volume, cu ft = ⅓ (3.14) (8 ft) (8 ft) (10 ft) = 670 cu ft
    Next find the volume of the cylindrical part of the tank.
    volume = πr² (depth) = (3.14) (8 ft) (8 ft) (30 ft) = 6,029 cu ft
    Then, add the two volumes for the answer.
    total volume, cu ft = 670 cu ft + 6,029 cu ft = 6,699 cu ft, round to 6,700 cu ft
    To find the number of gallons multiply the total number of cubic feet by 7.5 gal/cu ft.
    (6,700 cu ft) (7.5 gal/cu ft) = 50,250 gal, round to 50,000 gal

8. Answer: **b.** 38 in.

    Solution: First, calculate the number of acres in 45 square miles.
    (640 acres/sq mi) (45 miles) = 28,800 acres
    Next, determine the number of gallons per acre.

    $$\frac{(3{,}267 \text{ mil gal}) (1{,}000{,}00/1 \text{ M})}{28{,}800 \text{ acres}} = 113{,}438 \text{ gal/acre}$$

    Then find the thickness this water would have over the acre in inches.

    $$\frac{(113{,}438 \text{ gal/acre}) (12 \text{ in./ft})}{(43{,}560 \text{ sq ft/acre}) (7.5 \text{ gal/cu ft})} = 4.17 \text{ in.}$$

    (NOTE: This is only 11% of the rain.)

    Last, calculate the amount of rainfall in inches.
    (4.17 in.) (100%/11%) = 38 in.

9. Answer: **b.** 5.3 hours

    Solution: First, determine the number of gallons in the three flocculation basins, the sedimentation basin, and filters.
    The equation is: volume, gal = (length) (width) (depth) (7.5 gal/cu ft) (no. of basins or filters)
    volume, gal in floc basins = (48 ft) (20 ft) (12.5 ft) (7.5 gal/cu ft) (3 basins) = 270,000 gal
    volume, gal in sed basin = (565 ft) (72 ft) (10 ft) (7.5 gal/cu ft) = 3,051,000 gal, rounded to 3,000,000 gal

volume, gal in filters = (40 ft) (28 ft) (10.5) (7.5 gal/cu ft) (8 filters) = 705,600 gal, rounded to 700,000 gal

total = 270,000 gal + 3,000,000 gal + 700,000 gal = 3,970,000 gal

Next, convert million gallons per day to gallons per hour.

(18.1 mgd) (1 day/24 hours) (1,000,000/ 1M) = 754,167 gph, round to 754,000 gph

Write the equation with units asked for in the question.

$$\text{detention time, hr} = \frac{\text{volume, gal}}{\text{flow rate, gph}}$$

$$\text{detention time, hr} = \frac{3{,}970{,}00 \text{ gal}}{754{,}000 \text{ gph}} = 5.3 \text{ hours}$$

10. Answer: **b.** 11.5 psi

    Solution: The equation is: $\text{psi} = \dfrac{\text{depth, ft}}{2.31 \text{ ft/psi}}$

    $$\text{psi} = \frac{26.5 \text{ ft}}{2.31 \text{ ft/psi}} = 11.5 \text{ psi}$$

11. Answer: **a.** 1.23

    Solution: The density of water can also be expressed as lb/gal, or 8.34 lb/gal.

    $$\text{specific gravity} = \frac{10.27 \text{ lb/gal}}{8.34 \text{ lb/gal}} = 1.23 \text{ specific gravity}$$

12. Answer: **b.** 2.6 fps

    Solution: First, convert the number of gallons per minute to cubic feet per second.

    $$\text{no. of cfs} = \frac{60 \text{ gpm}}{(7.5 \text{ gal/cu ft})(60 \text{ sec/min})} = 0.13 \text{ cfs}$$

    Next, convert the diameter from inches to feet.

    no. of ft = (3 in.) (1 ft/12 in.) = 0.25 ft

    Equation: flow, cfs = (area, sq ft) (velocity, fps); where the area = $(0.785)(D)^2$

    0.13 cfs = (0.785) (0.25 ft) (0.25 ft) (flow, fps)

    Rearrange and solve for the flow in feet per second.

    $$\text{flow, fps} = \frac{0.13 \text{ cfs}}{(0.785)(0.25)(0.25)} = 2.6 \text{ fps}$$

118 CERTIFICATION STUDY GUIDE

13. Answer: **b.** 3.05 mg/L

    Solution: The formula is: chlorine dose = chlorine demand + chlorine residual

    Rearrange and solve for chlorine demand.

    chlorine demand = chlorine dose − chlorine residual

    chlorine demand = 4.25 mg/L − 1.20 mg/L = 3.05 mg/L

14. Answer: **c.** 1,700 lb/day

    Solution: Convert the percent purity to decimal form. 49%/100% = 0.49

    Formula is: lb/day = $\dfrac{(mgd)(dosage, mg/L)(8.34\ lb/gal)}{(\%\ purity)}$

    lb/day, alum = $\dfrac{(8.2\ mgd)(12\ mg/L)(8.34\ lb/gal)}{(0.49\ purity)}$

    lb/day, alum = 1,674.8 lb/day, round to 1,700 lb/day of alum

15. Answer: **a.** 22 mgd

    Solution: First, find the total chlorine dosage.
    Equation is: total Cl_2 dosage = Cl_2 demand + Cl_2 residual
    total Cl_2 dosage = 2.3 mg/L (demand) + 1.2 mg/L (residual) = 3.5 mg/L
    Then use the "pounds" formula, but solve for the unknown, million gallons per day.

    lb/day = (mgd)(dosage, mg/L)(8.34 lb/gal)

    mgd = $\dfrac{lb/day}{(dosage)(8.34\ lb/gal)}$

    mgd = $\dfrac{650\ lb/day}{(3.5\ mg/L)(8.34\ lb/gal)}$ = 22.3 mgd, round to 22 mgd

16. Answer: **b.** 50 oz

    Solution: First, solve the problem using the modified "pounds" formula.

    Write the equation: lb = $\dfrac{(mil\ gal)(dosage, mg/L)(8.34\ lb/gal)}{\%\ purity}$

    Again, delete "day" on each side of the equation because it is not needed. Next, find how many million gallons there are in 835 gal.

    mil gal = $\dfrac{835\ gal}{1,000,000,000/1\ M}$ = 0.000835 mil gal

    Substitution: lb of hypo =

    $\dfrac{(0.000835\ mil\ gal)(50\ mg/L)(8.34\ lb/gal)}{12\%/100\%}$ = 2.90 lb, round to 3 lb of hypo

WATER TREATMENT   119

Now convert pounds to gallons.

$$\frac{3 \text{ lb of hypo}}{8.34 \text{ lb/gal}} = 0.4 \text{ gal}$$

Now convert gallons to ounces.

no. of ounces, hypo = (128 oz/gal) (0.4 gal) = 51.2 oz, round to 50 oz of hypo

17. Answer: **d.** 46 gpm

   Solution: First, determine the number of minutes the pump was working.

   3 hr (60 min/hr) + 45 min = 180 min + 45 min = 225 min

   Then determine the number of gallons per minute by dividing the number of gallons pumped by the total time the pump worked.

   $$\frac{10,350 \text{ gal}}{225 \text{ min}} = 46.0 \text{ gpm}$$

18. Answer: **d.** 68 mhp

   Solution: First, convert million gallons per day to gallons per minute.

   gpm = (1.35 mgd) (1,000,000/1 M) (1 day/1,440 min) = 937.5 gpm, round to 938 gpm

   The equation is: mhp = $\dfrac{(\text{flow, gpm})(\text{TH, ft})}{(3,960)(\text{motor effic.})(\text{pump effic.})}$

   The 3,960 is a constant. Next, convert the percent motor and pump efficiencies to decimal form by dividing by 100%.

   motor efficiency = 90%/100% = 0.90
   pump efficiency = 78%/100% = 0.78
   The equation is: mhp =

   $$\frac{(938 \text{ gpm})(202 \text{ ft})}{(3,960)(0.90 \text{ motor effic.})(0.78 \text{ pump effic.})} = 68 \text{ mhp}$$

19. Answer: **c.** 59 cfs

   Solution: Equation is: Q (flow) = (area) (velocity)
   Q, cfs = (7 ft) (3 ft) (2.8 fps) = 59 cfs

20. Answer: **d.** 17,000 gpd/ft

   Solution: The equation is: weir overflow rate = $\dfrac{\text{flow, gpd}}{\text{weir length, ft}}$

   First change 1.7 mgd to gallons per day: 1.7 mgd (1,000,000/1 M) = 1,700,000 gpd

$$\text{weir overflow rate} = \frac{1{,}700{,}000 \text{ gpd}}{100 \text{ ft}} = 17{,}000 \text{ gpd/ft}$$

21. Answer: **d.** 590 gpd/sq ft

    Solution: First, convert the number of cubic feet per second to gallons per day.

    (17 cfs)(86,400 sec/day)(7.5 gal/cu ft) =
    11,016,000 gal/day, round to 11,000,000 gal/day

    The equation for surface loading rate is:

    $$\text{surface loading rate} = \frac{\text{gallons per day (gpd)}}{\text{no. of sq ft}}$$

    $$\text{surface loading rate} = \frac{11{,}000{,}000 \text{ gpd}}{(375 \text{ ft})(50 \text{ ft})} =$$
    586.67 gpd/sq ft, round to 590 gpd/sq ft

22. Answer: **b.** 18 ft

    Solution: Write the equation, arranging it to solve for the unknown, drawdown.

    $$\text{drawdown, ft} = \frac{\text{well yield, gpm}}{\text{specific yield, gpm/ft}}$$

    $$\text{drawdown, ft} = \frac{320 \text{ gpm}}{18.2 \text{ gpm/ft}} = 17.6 \text{ ft, round to 18 ft}$$

23. Answer: **d.** 57.75 ft

    Solution: Equation: drawdown, ft = pumping water level, ft − static water level, ft.
    Rearrange the equation to solve for pumping water level.
    pumping water level, ft = drawdown, ft + static water level
    Substitute known values.
    pumping water level, ft = 19.52 ft + 38.23 ft = 57.75 ft

24. Answer: **c.** 174 mg/L as $CaCO_3$

    Solution: The equation is: total hardness, mg/L as $CaCO_3$ = Ca hardness, mg/L as $CaCO_3$ + Mg hardness, mg/L as $CaCO_3$
    Simple substitution: total hardness, mg/L as $CaCO_3$ =
    120 mg/L Ca + 54 mg/L Mg = 174 mg/L as $CaCO_3$

WATER TREATMENT 121

25. Answer: **b.** 66 ft

   Solution: Write the equation: total head, ft = total static head, ft (difference in elevation) + head losses, ft

   total head, ft = (232 ft − 175 ft) + 9 ft = 57 ft + 9 ft = 66 ft

26. Answer: **a.** 2.26 mg/L of chlorine

   Solution: Equation is: lb/day = (mgd) (dosage, mg/L) (8.34 lb/gal)
   Substituting and rearranging the equation:

   $$\frac{485 \text{ lb/day}}{(25.7 \text{ mgd})(8.34 \text{ lb/gal})} = \text{dosage, mg/L}$$

   dosage, mg/L = 2.26 mg/L of chlorine

27. Answer: **d.** 14,000 cu ft

   Solution: The volume formula for a circular tank is: volume = πr^2 (height), where r is the radius.

   volume of tank in cu ft = 3.14 (15 ft) (15 ft) (20 ft) = 14,130 cu ft, round to 14,000 cu ft

28. Answer: **b.** 1.5 ft

   Solution: Equation: Q (flow) = (area) (velocity)
   12.8 cfs = (4.5 ft) (x ft, depth) (1.9 fps)
   Now solve for depth.

   $$x \text{ ft, depth} = \frac{12.8 \text{ cfs}}{(4.5 \text{ ft})(1.9 \text{ fps})} = 1.5 \text{ ft deep}$$

29. Answer: **b.** 9.6 hours

   Solution: First determine the volume in gallons for the clarifier.
   volume, gal = (0.785) (diameter)^2 (depth) (7.5 gal/cu ft)
   volume, gal = (0.785) (160.0 ft) (160.0 ft) (10.25 ft) (7.5 gal/cu ft) = 1,544,800 gal, round to 1,545,000 gal

   Then, convert million gallons per day to gallons per hour because detention time is asked for in hours.
   (3.86 mgd) (1,000,000/1 M) (1 day/24 hr) = 160,833 gpm, round to 161,000 gpm

   Equation is: detention time, hr = $\frac{\text{volume, gal}}{\text{flow rate, gph}}$

   detention time, hr = $\frac{1,545,000 \text{ gal}}{161,000 \text{ gph}}$ = 9.6 hours

30. Answer: **d.** 16,000 ppm alum

   Solution: If a 1% solution has 10,000 ppm, a 1.6% will have:
   (1.6%) (10,000 ppm/1%) = 16,000 ppm alum

122 CERTIFICATION STUDY GUIDE

31. Answer: **b.** 8.0 mg/L

    Solution: First, convert pounds per million gallon by dividing by 8.34 lb/gal.

    $$\text{lb/mil lb} = \frac{(67 \text{ lb})(1 \text{ gal})}{(1 \text{ mil gal})(8.34 \text{ lb})} = \frac{67 \text{ lb, gal}}{8.34 \text{ lb, mil gal}}$$

    Next, cancel out like units, lb and gal, to get: $\frac{8}{M}$ = 8.0 mg/L

32. Answer: **d.** 15,045 gal

    Solution: Equation is: Volume = (0.785) (D)² (height)
    volume = (0.785) (12.8 ft) (12.8 ft) (15.6 ft) = 2,006 cu ft
    Next, find the number of gallons.
    (2,006 cu ft) (7.5 gal/cu ft) = 15,045 gal

33. Answer: **a.** 18.5 hours

    Solution: First determine the volume in gallons for the sed basin.
    volume, gal = (295 ft) (80.0 ft) (11.0 ft) (7.5 gal/cu ft) = 1,947,000 gal
    Then, find the volume of the filters.
    volume, gal = (40.0 ft) (30.0 ft) (7.00 ft) (12.0 filters) (7.5 gal/cu ft) = 756,000 gal
    The total volume would then be the sum of the filters and sed basin.
    total volume, gal = 1,947,000 + 756,000 = 2,703,000 gal
    Next, convert million gallons per day to gallons per hour as asked for in problem.
    (3.50 mgd) (1,000,000/1 M) (1 day/24 hr) = 145,833 gph, round to 146,000 gph

    Equation is: detention time, hr = $\frac{\text{volume, gal}}{\text{flow rate, gph}}$

    detention time, hr = $\frac{2,703,000 \text{ gal}}{146,000 \text{ gph}}$ = 18.5 hours

34. Answer: **c.** 45 mL/min of polymer

    Solution: Find the number of pounds per day of polymer required by using the "pounds" equation.
    lb/day, polymer = (mgd) (dosage, mg/L) (8.34 lb/gal)
    lb/day, polymer = (14.3 mgd) (1.5 mg/L) (8.34 lb/gal) = 179 lb/day
    Next, determine the pounds per gallon of the polymer solution.
    lb/gal = (specific gravity) (8.34 lb/gal) = (1.23)(8.34 lb/gal) = 10.3 lb/gal
    Convert the number of pounds per day to number of gallons per day.

$$\text{gpd, polymer} = \frac{179 \text{ lb/day}}{10.3 \text{ lb/gal}} = 17 \text{ gpd}$$

Then convert gallons per day to milliliters per minute.

mL/min of polymer =

$$\frac{(17 \text{ gpd})(3{,}785 \text{ mL/gal})}{1{,}440 \text{ min/day}} = 45 \text{ mL/min of polymer}$$

35. Answer: **b.** 32 lb/day

    Solution: First determine how many million gallons per day is being treated.

    mgd = (1,750 gpm)(1,440 min/day)(1 M/1,000,000) = 2.52 mgd

    Since there is natural fluoride (F) present, subtract the natural from the desired to get the dose required.

    F dose required = 1.05 mg/L F – 0.15 mg/L natural F content = 0.90 mg/L F

    Write the "pounds" equation with the addition of the percent purity and fluoride (F) content.

    $$\text{lb/day, Na}_2\text{SiF}_6 \text{ compound} = \frac{(\text{mgd})(\text{dosage, mgL})(8.34 \text{ lb/day})}{(\% \text{ purity}/100\%)(\% \text{ F content}/100\%)}$$

    lb/day, Na$_2$SiF$_6$ compound =

    $$\frac{(2.52 \text{ mgd})(0.90 \text{ mg/L})(8.34 \text{ lb/day})}{(98\% \text{ purity}/100\%)(60.6\% \text{ F content}/100\%)} = 32 \text{ lb/day}$$

36. Answer: **b.** 3.6 lb

    Solution: First, find the length (in feet) of water-filled casing.

    length of water-filled casing = depth of well – depth of water to top of casing

    length of water-filled casing = 328 ft – 60 ft = 268 ft

    Then, determine the volume in gallons of water in the well casing using the following formula:

    volume, in gal = (0.785)(D)^2 (length)(7.5 gal/cu ft)

    volume, in gal = (0.785)(1.5 ft)(1.5 ft)(268 ft)(7.5 gal/cu ft) = 3,550 gal

    Next, determine the number of million gallons.

    mil gal = (3,550 gal)(1 M/1,000,000) = 0.00355 mil gal

    Last, using the "pounds" equation, calculate the number of pounds of calcium hypochlorite.

    $$\text{lb of calcium hypochlorite} = \frac{(0.00355 \text{ mil gal})(75 \text{ mg/L})(8.34 \text{ lb/gal})}{(62\%/100\% \text{ available chlorine})}$$

    lb of calcium hypochlorite = 3.6 lb

124 CERTIFICATION STUDY GUIDE

37. Answer: **b.** 11 gpm/sq ft

   Solution: Write the equation: BW rate, gpm/sq ft =

   $$\frac{\text{backwash pumping rate, gpm}}{\text{filter area, sq ft}}$$

   First, convert the pumping rate in cubic feet per second into gallons per minute.

   gpm = (20 cfs) (7.5 gal/cu ft) (60 sec/min) = 9,000 gpm

   BW rate, gpm/sq ft =

   $$\frac{9{,}000 \text{ gpm}}{840 \text{ sq ft}} = 10.7 \text{ gpm/sq ft, round to } 11 \text{ gpm/sq ft}$$

38. Answer: **d.** 2,000 lb/yr

   Solution: First, calculate the manganese removal in parts per million.
   (0.15 ppm) (88%/100%) = 0.15 ppm (0.88) = 0.132 ppm, round to 0.13 ppm
   Determine the amount of water in million gallons produced for the year.
   (5 mgd) (365 days/year) = 1,825 mil gal/yr
   Next, using the "pounds" equation, solve for the number of pounds per year.
   lb/yr = (mil gal/yr) (Dosage, mg/L) (8.34 lb/gal)
   lb/yr = (1,825 mil gal/yr) (0.13 mg/L) (8.34 lb/gal) = 1,979 lb/yr, round to 2,000 lb/yr of Mn removed

39. Answer: **b.** 2.0 fps

   Solution: Flow in 8-in. pipe equals the flow in the 12-in. pipe because the flow must remain constant.
   $Q_1 = Q_2$
   Since Q, flow = (area) (velocity), it follows that:
   (area 1) (velocity 1) = (area 2) (velocity 2)
   First, find the diameters in feet for the 8-in. and 12-in. pipes.
   diameter for 8-in. = 8-in. (1 ft/12-in.) = 0.667 ft
   diameter for 12-in. = 12-in. (1 ft/12-in.) = 1.0 ft
   Then determine the areas of each size pipe.
   area = $(0.785)(D)^2$
   area 1 (8-in.) = (0.785) (0.667 ft) (0.667 ft) = 0.35 sq ft
   area 2 (12-in.) = (0.785) (1.0 ft) (1.0 ft) = 0.785 sq ft
   Last, substitute areas calculated and known velocity in 8-in. pipe.
   (0.35 sq ft) (4.5 fps) = (0.785 sq ft) (x, fps)

Solve for x:

$$x, \text{fps} = \frac{(0.35 \text{ sq ft})(4.5 \text{ fps})}{(0.785 \text{ sq ft})} = 2.0 \text{ fps in 12-in. pipe}$$

40. Answer: **c.** 3.5 psi

    Solution: 1 psi = 2.31 ft

    Thus: $\dfrac{(\text{no. of ft})(1 \text{ psi})}{(2.31 \text{ ft})}$

    Substitution: $\dfrac{(8.0 \text{ ft})(1 \text{ psi})}{(2.31 \text{ ft})} = 3.5 \text{ psi}$

    Or alternatively,
    0.433 psi/ft
    Thus: (0433 psi/ft) (8.0 ft) = 3.5 psi

41. Answer: **d.** 21,000 lb/month of lime

    Solution: First convert gallons per minute to million gallons per day.
    (2,850 gpm) (1,440 min/day) (1 M/1,000,000) = 4.10 mgd

    Next convert grams per minute of lime to pounds per day.
    (220 g/min) (1 lb/454 g) (1,440 min/day) =
      698 lb/day, round to 700 lb/day
    (700 lb/day) (30 days/month) = 21,000 lb/month of lime

42. Answer: **a.** 1,155 lb/day

    Solution: lb/day of soda ash = (g/min) (1,440 min/day) (1 lb/454 g) = lb/day
    lb/day of soda ash = (364 g/min) (1,440 min/day) (1 lb/454 g) =
      1,155 lb/day of soda ash

43. Answer: **a.** 8%

    Solution: A 1% solution = 10,000 ppm

    $\dfrac{80,350 \text{ ppm}}{10,000 \text{ ppm}/1\%} = 8.035\%$, round to 8%

44. Answer: **a.** 0.5: A positive Langelier Index indicates the water is scale forming.

    Solution: Write the equation: Langelier Index (L.I.) = pH − pH$_s$
    L.I. = 8.0 − 7.5 = 0.5 A positive L.I. indicates the water is scale forming

45. Answer: **c.** 8 in.

    Solution: Flow, cfs = (area, sq ft) (velocity, fps); where the area = (0.785) (D)^2
    1.6 cfs = (0.785) (D)^2 (4.6 fps)

Rearrange and solve for the diameter.

$$D^2 = \frac{1.6 \text{ cfs}}{(0.785)(4.6 \text{ fps})} \quad D^2 = 0.443 \text{ ft} \quad D = 0.67 \text{ ft}$$

$$D = (0.67)(12 \text{ in./ft}) = 8 \text{ in.}$$

46. Answer: **b.** 6.5 psi

    Solution: First calculate the number of cubic feet of water present.

    $$\frac{140,900 \text{ gal}}{7.5 \text{ gal/cu ft}} = 18,787 \text{ cu ft}$$

    The number of cubic feet equals πr^2 (depth).
    Solve for depth.
    18,837 cu ft = 3.14 (20 ft) (20 ft) (depth).

    $$\text{depth} = \frac{18,787 \text{ cu ft}}{3.14(20 \text{ ft})(20 \text{ ft})}$$

    depth = 15 ft

    Now solve for the number of pounds per square inch at the bottom of the tank.

    $$\text{psi} = \frac{\text{depth}}{2.31 \text{ ft/psi}} = \frac{15 \text{ ft}}{2.31 \text{ ft/psi}} = 6.5 \text{ psi}$$

47. Answer: **b.** 4.3 psi

    Solution: First calculate the number of cubic feet of water present.

    $$\frac{140,900 \text{ gal}}{7.5 \text{ gal/cu ft}} = 18,787 \text{ cu ft}$$

    The number of cubic feet equals πr^2 (depth).

    Solve for depth.

    $$\text{depth} = \frac{18,787 \text{ cu ft}}{3.14 (20 \text{ ft}) (20 \text{ ft})}$$

    depth = 15 ft

    Now solve for the number of pounds per square inch at the bottom of the tank.

    $$\text{psi} = \frac{\text{depth}}{2.31 \text{ ft/psi}} = \frac{15 \text{ ft}}{2.31 \text{ ft/psi}} = 6.5 \text{ psi}$$

Finally, solve for pounds per square inch 5 ft above the bottom of the tank. The depth of the water is 5 ft less than the total depth (15 ft – 5 ft = 10 ft). Thus:

$$\text{psi} = \frac{\text{depth}}{2.31 \text{ ft/psi}} = \frac{10 \text{ ft}}{2.31 \text{ ft/psi}} = 4.3 \text{ psi}$$

48. Answer: **b.** 12.1 hours

    Solution: Write the equation with units asked for in question.

    $$\text{detention time, hr} = \frac{\text{volume, gal}}{\text{flow rate, gph}}$$

    Next, determine the capacity in gallons for each basin by converting from million gallons to gallons for the clear well. Then add all basins for total volume in gallons.

    volume, gal floc basins =
      (45.0 ft) (9.0 ft) (10.0 ft) (7.5 gal/cu ft) (5 basins)   =   151,875 gal
    volume of sed. basin =
      (675 ft) (45.0 ft) (10.0 ft) (7.5 gal/cu ft)   =   2,278,125 gal
    volume of filters =
      (35.0 ft) (25.0 ft) (12.0 ft) (7.5 gal/cu ft) (8 filters)   =   630,000 gal
    volume of clear well =
      (2.8 mil gal) (1,000,000/1 M)   =   2,800,000 gal
                                                                5,860,000 gal

    Then, convert the flow rate of 11.6 mgd to gallons per hour.

    $$\text{gph} = (11.6 \text{ mgd}) \frac{(1,000,000 \text{ gal})}{1 \text{M}} \frac{(1 \text{ day})}{24 \text{ hr}} = 483,333 \text{ gph}$$

    Substitution in equation above:

    $$\text{detention time, hr} = \frac{5,860,000 \text{ gal}}{483,333 \text{ gph}} = 12.1 \text{ hours}$$

49. Answer: **d.** 1.7 %

    Solution: First convert 7,500 gpm filter flow to million gallons per day.

    $$(7,5000 \text{ gpm}) \left(1,440 \frac{\text{min}}{\text{day}}\right) \left(\frac{1 \text{ mil gal}}{1,000,000 \text{ gal}}\right) = 10.8 \text{ mgd, round to 11 mgd}$$

    Then convert 1,750 gpd of the hypochlorite solution rate of flow to million gallons per day.

    $$(1,750 \text{ gpd}) \left(\frac{1 \text{ mil gal}}{1,000,000 \text{ gal}}\right) = 0.00175 \text{ mgd}$$

Then, using the equal dosage equations:

$$(0.00175 \text{ mgd})(x \text{ mg/L})(8.34 \text{ lb/gal}) = (11 \text{ mgd})(2.75 \text{ mg/L})(8.34 \text{ lb/gal})$$

$$x, \text{mg/L} = \frac{(11 \text{ mgd})(2.75 \text{ mg/L})(8.34 \text{ lb/gal})}{(0.00175 \text{ mgd})(8.34 \text{ lb/gal})} =$$

17,286 mg/L, round to 17,000 mg/L

Last, convert milligram per liter into percent.

$$(17,000 \text{ mg/L}) \frac{(1\%)}{10,000} = 1.7\%$$

50. Answer: **a.** 13.5 billion lb/yr

    Solution: Write the equation: lb/day =
    (mgd) (concentration, mg/L) (8.34 lb/gal) (removal efficiency)
    lb/day of salts removed =
    (1,500) (2,978 mg/L) (8.34 lb/gal) (99.3%/100%) = 36,993,997 lb/day,
    round to 37,000,000 lb/day

    Then, find the pounds per year removed.
    (37,000,000 lb/day) (365 days/yr) = 13,505,000,000 lb/yr

DISTRIBUTION OPERATOR
CERTIFICATION EXAMS

SYSTEM DESIGN

System Design	Class I	Class II	Class III	Class IV
Assess system demand	Application	Application	Analysis	Analysis
Design joint restraints	Application	Application	Analysis	Analysis
Design shoring	Application	Application	Analysis	Analysis
Design thrust blocks	Application	Application	Analysis	Analysis
Flushing program	Application	Application	Application	Application
System layout	Application	Application	Analysis	Analysis
System map	Application	Application	Analysis	Analysis
Perform pressure readings	Application	Application	Analysis	Analysis
Read blueprints, readings, and maps	Application	Application	Analysis	Analysis
Select materials	Application	Application	Analysis	Analysis
Select type of pipes	Application	Application	Analysis	Analysis
Size mains	Application	Application	Analysis	Analysis
Write plans (i.e. operations and maintenance plans)	Application	Application	Analysis	Analysis

Suggestions for Study:
- Ability to adjust equipment
- Ability to diagnose/troubleshoot system units
- Ability to discriminate between normal and abnormal conditions
- Ability to monitor electrical and mechanical equipment
- Knowledge of blueprint readings
- Knowledge of cathodic protection
- Knowledge of different types of joints, restraints and thrust blocks
- Knowledge of general electrical, mechanical and hydraulic principles
- Knowledge of fireflow requirements
- Knowledge of measuring instruments
- Knowledge of piping material, type and size
- Knowledge of pneumatics
- Knowledge of regulations
- Knowledge of start-up and shut down procedures
- Knowledge of testing instruments

SYSTEM DESIGN

Sample Questions for Class I, answers on p. 197

1. The size of water mains, pumping stations, and storage tanks is primarily determined by
   a. Maximum day demand during any 24-hour period during the previous year
   b. Population served
   c. Per-capita water use
   d. Fire protection requirements

2. Which of the following is a measure of the smoothness of a pipe?
   a. S factor
   b. C value
   c. Hazen-Williams formula
   d. T factor

3. Why is excessive water pressure to residential homes objectionable?
   a. Increases particulate matter reaching the customer
   b. Causes erosion of the copper plumbing due to the high velocities, giving the water a metallic taste
   c. Decreases the life of water heaters and other water-using appliances
   d. Causes foaming from faucets

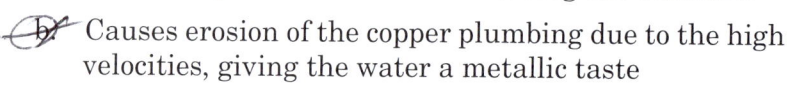

4. What is the best time to perform a flushing program on the mains?
   a. Spring when the weather is usually mild
   b. Fall when water usage is low and the weather not yet harsh
   c. Late at night to lessen traffic disruption and minimize customer complaints
   d. Summer when many residents are on vacation

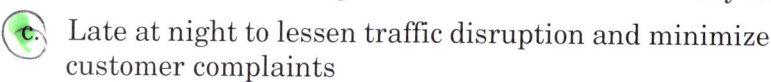

5. Calculate the pounds per square inch pressure at the bottom of a tank, if the water level is 33.11 ft deep?
   a. 14.3 psi
   b. 28.6 psi
   c. 33.1 psi
   d. 76.5 psi

SYSTEM DESIGN

Sample Class II Questions, answers on p. 198

1. What type of ductile-iron pipe joint is used primarily for river crossings and underwater intakes, and occasionally in rough terrain?

    a. Restrained

    b. Mechanical

    c. Flanged

    d. Ball-and-socket

2. In general, a trench should be no more than how many feet wider than the diameter of the pipe?

    a. 1 to 2 ft

    b. 2 to 3 ft

    c. 3 to 4 ft

    d. 4 to 5 ft

3. Which of the following is a distribution layout pattern?

    a. Dendritic

    b. Grid

    c. Parallel

    d. Brush

4. What is the cross-sectional area of a pipe that is 10 in. in diameter?

    a. 0.24 sq ft

    b. 0.54 sq ft

    c. 0.65 sq ft

    d. 0.79 sq ft

5. The utility's annual average day demand is determined by dividing the

    a. Water use per month by the number of days in that month

    b. Total water use for a year by 365 days

    c. Water use per week by seven days

    d. Average day demand by the number of residents

SYSTEM DESIGN

Sample Class III Questions, answers on p. 199

1. If flow through a 10-in. pipe is 722 gpm, what is the velocity in feet per second?

    a. 2 fps

    b. 3 fps

    c. 5 fps

    d. 8 fps

2. Complaints of poor water quality will most likely occur in which type of water main layout?

    a. Arterial loop

    b. Tree

    c. Grid

    d. Nebular

3. A main break may cause loss of pressure in the distribution system, which in turn may result in

    a. Contamination of the system by backsiphonage

    b. "Ice" formation in the pipes

    c. Increase in chlorine residual

    d. Water hammer

4. Vacuum breakers are designed to be used on piping connections where

    a. System pressure may be less than atmospheric

    b. Backpressure will not exist

    c. System pressure may be less than customer's device pressure but not less than atmospheric

    d. Backpressure will occur

5. What is the recommended minimum water pressure in a distribution system at any time, including fire flow conditions?

    a. 20 psi

    b. 25 psi

    c. 30 psi

    d. 35 psi

SYSTEM DESIGN

Sample Questions for Class IV, answers on p. 200

1. Determine the volume of water in gallons for the following distribution system:

    - Distribution pipe "A" is 1,376 ft in length and 3.0 ft in diameter.
    - Distribution pipe "B" is 833 ft in length and 2.0 ft in diameter.
    - Storage tank is 120 ft in diameter with a water height of 30.73 ft.

    a. 73,000 gal
    b. 93,000 gal
    c. 2,700,000 gal
    d. 5,400,000 gal

2. Which of the following will least likely increase soil corrosiveness?

    a. High acidity
    b. Low soil moisture content
    c. High alkalinity
    d. Presence of sulfide

3. What type of pipe joint is best to use under high groundwater and muddy conditions despite best efforts to pump the groundwater out of the trench?

    a. Push joint
    b. Mechanical joint
    c. Boltless ball joint
    d. Tied rubber gasket joint

4. An arterial-loop distribution system has flow from how many directions?

    a. 1
    b. 2
    c. 3
    d. 4

5. What should the flow meter read in gallons per minute if a 12-in. diameter main is to be flushed at 5.0 fps?

    a. 39.2
    b. 78.5
    c. 392
    d. 1,764

MONITOR WATER QUALITY

Monitor, Evaluate and Adjust Disinfection	Class I	Class II	Class III	Class IV
Chlorine disinfection	Application	Application	Analysis	Analysis
Water Quality				
Cross connection surveys/control	Application	Application	Application	Analysis
Sample site plan	Application	Analysis	Analysis	Analysis
Sanitary surveys	Analysis	Analysis	Analysis	Analysis
Inspect springs	Comprehension	Comprehension	Comprehension	Comprehension
Inspect surface water	Comprehension	Comprehension	Comprehension	Comprehension
Inspect wells	Analysis	Analysis	Analysis	Analysis
Water Quality Parameters and Sampling				
Chlorine demand/residual/dosage	Application	Application	Analysis	Analysis
Coliforms	Application	Application	Analysis	Analysis
Conductivity	Application	Application	Application	Application
Lead/copper	Application	Application	Analysis	Analysis
pH	Application	Application	Analysis	Analysis
Temperature	Application	Application	Analysis	Analysis
Turbidity	Application	Application	Analysis	Analysis

Suggestions for Study:

- Ability to adjust flow patterns and system units
- Ability to calibrate instruments
- Ability to communicate verbally and in writing
- Ability to diagnose/troubleshoot system units
- Ability to discriminate between normal and abnormal conditions
- Ability to evaluate system units
- Ability to follow written procedures
- Ability to inspect pumps
- Ability to interpret Material Safety Data Sheets
- Ability to maintain system in normal operating condition
- Ability to perform basic math
- Ability to recognize normal and abnormal analytical results
- Knowledge of disinfection concepts and design parameters
- Knowledge of disinfection process
- Knowledge of general chemistry, biology & physical science
- Knowledge of general electrical and hydraulic principles
- Knowledge of hydrology

Suggestions for Study: (continued)

- Knowledge of laboratory equipment
- Knowledge of monitoring requirements
- Knowledge of normal characteristics of water
- Knowledge of principles of measurement
- Knowledge of proper chemical handling and storage
- Knowledge of proper sampling procedures and requirements
- Knowledge of public notification requirements
- Knowledge of quality control/quality assurance practices
- Knowledge of regulations and standards
- Knowledge of reporting requirements
- Knowledge of Safe Drinking Water Act
- Knowledge of safety procedures
- Knowledge of sanitary survey process
- Knowledge of well drilling principles
- Knowledge of well-head protection

MONITOR WATER QUALITY

Sample Questions for Class I, answers on p. 201

1. As the temperature of the water increases, the disinfecting action of chlorine is
   a. More effective
   b. Not affected
   c. Less effective
   d. Likely to increase DO

2. Large full storage tanks are typically disinfected with a chlorine residual of
   a. 3 mg/L
   b. 10 mg/L
   c. 25 mg/L
   d. 50 mg/L

3. Why is a well acidified?
   a. Take out soluble iron or manganese
   b. Increase the well's productivity
   c. Remove objectionable gases
   d. Remove turbidity

4. What is the **primary** source of coliforms in a water supply?
   a. Insects that live or die in the water source
   b. Protozoans and other microorganisms
   c. Fertilizers
   d. Fecal material from warm-blooded animals

5. pH is a measure of
   a. Conductivity
   b. Water's ability to neutralize acid
   c. Hydrogen ions
   d. Dissolved solids

MONITOR WATER QUALITY

Sample Questions for Class II, answers on p. 202

1. What is the recommended minimum contact time when disinfecting water mains with the chlorine slug method?

    a. 3 hours
    b. 6 hours
    c. 10 hours
    d. 12 hours

2. Which of the following best defines the term *static water level*?

    a. Water level in a well after a pump has operated for a period of time
    b. Water level in a well when the well is not in operation
    c. Water level in a well measured from the ground surface to the drawdown water level
    d. Water level in a well measured from the natural water level to the drawdown water level

3. Where are chlorine samples typically collected from in the distribution system?

    a. At points representative of conditions within the system
    b. Uniformly distributed throughout the system as much as possible
    c. At the extreme locations of the system
    d. At representative points throughout the system based on elevation

4. Which of the following best defines the cone of depression?

    a. Change in water elevation from the normal level to the pumping level
    b. Depression around the well of the water surface caused by pumping water from the well
    c. Water level in a well after a pump has operated over a period of time
    d. Measured distance from the ground to the pumping level

5. Turbidity is caused by

    a. Dissolved solids
    b. Suspended particles
    c. Dissolved gases
    d. Dissolved colored solids

MONITOR WATER QUALITY

Sample Class III Questions, answers on p. 203

1. What is the term for water samples collected at regular intervals and combined in equal volume with each other?

    a. Time grab samples

    b. Time flow samples

    c. Time composite samples

    d. Proportional time composite samples

2. What chemical is present in a bacteria sample bottle for the purpose of neutralizing chlorine?

    a. Sodium benzoate

    b. Sodium thiosulfate

    c. Sodium phenoxide

    d. Sodium salicylate

3. Chlorine neutralization is necessary when a treated water sample is to be analyzed for

    a. Iron

    b. Bacteria

    c. Manganese

    d. Nitrate

4. The residual drawdown of a well is defined as

    a. Water level in a well after a pump has operated over a period of time

    b. Measured distance from the ground to the pumping level

    c. Water level below the normal level that persists after a well pump has been off for a period of time

    d. Measured distance between the water level and the top of the screen

5. What is the basis for the number of samples that must be collected for utilities monitoring for lead and copper that are in compliance or have installed corrosion control?

    a. Size of distribution system

    b. Population

    c. Amount of water produced

    d. Number of raw water sources

MONITOR WATER QUALITY

Sample Class IV Questions, answers on p. 204

1. Where should bacteriological samples be collected in the distribution system?

    a. Uniformly distributed throughout the system based on area

    b. At locations that are representative of conditions within the system

    c. Almost always from extreme locations in the system but occasionally at other locations

    d. Uniformly throughout the system based on population density

2. The quantity of oxygen that can remain dissolved in water is related to

    a. Temperature

    b. pH

    c. Turbidity

    d. Alkalinity

3. In coliform analyses using the presence-absence test, a sample should be incubated for

    a. 24 hours at 25°C

    b. 36 hours at 35°C

    c. 24 and 36 hours at 25°C

    d. 24 and 48 hours at 35°C

4. What is the potential cross-connection hazard from sprinkler systems?

    a. Low

    b. Low to moderate

    c. Moderate

    d. High

5. A well produces 365 gpm with a drawdown of 22.5 ft. What is the specific yield in gallons per minute per foot?

    a. 16.2

    b. 22.5

    c. 32.4

    d. 36.5

INSTALL UNITS

Install Units	Class I	Class II	Class III	Class IV
Backflow prevention devices	Comprehension	Comprehension	Application	Application
Hydrants	Comprehension	Application	Application	Analysis
Meters	Application	Application	Application	Analysis
Piping and fitting	Application	Application	Application	Analysis
Rigging	Application	Application	Application	Application
Service connections	Application	Application	Application	Analysis
Shoring	Application	Application	Application	Analysis
Taps	Application	Application	Application	Analysis
Valves	Application	Application	Application	Analysis
Water mains	Application	Application	Application	Analysis
Water storage facility	Comprehension	Comprehension	Comprehension	Application

Suggestions for Study:

- Ability to follow written procedures
- Knowledge of approved backflow prevention devices
- Knowledge of facility operation and maintenance
- Knowledge of function of tools
- Knowledge of pipe fittings and joining methods
- Knowledge of piping material, type and size
- Knowledge of regulations
- Knowledge of start-up and shut-down procedures

INSTALL UNITS

Sample Class I Questions, answers on p. 205

1. What is the term for a framework of wood or metal installed to prevent caving of trench walls?
    a. Sheeting
    b. Sloping
    c. Shielding
    d. Shoring *(circled)*

2. Which one of the following service line materials is **not** flexible?
    a. Copper
    b. Galvanized iron *(circled)*
    c. Lead
    d. Plastic

3. Which type of pipe is typically used in the construction of very large water mains?
    a. Asbestos-cement
    b. Steel *(circled)*
    c. Plastic
    d. Galvanized

4. Air-relief valves are installed to
    a. Release some of the energy created by water hammer
    b. Stop flow completely when the tank is full
    c. Vent air that has accumulated in the well column while the well is not in use *(circled)*
    d. Turn off pressure during hydrant main maintenance without disrupting service to customers

5. After a new water main is installed and pressure tested, it should be
    a. Flushed with clean water for 24 hours and put into service
    b. Filled with a solution of 25 ppm to 50 ppm free chlorine for at least 24 hours prior to flushing *(circled)*
    c. Filled with clean water and allowed to sit for 5 days at full pressure before turning the water into the system
    d. Photographed so that mapping can be avoided until the system is complete

INSTALL UNITS

Sample Class II Questions, answers on p. 206

1. A 2-in. tap can be made through a saddle on which of the following ductile iron pipe sizes?

    a. 6-in.
    b. 8-in.
    c. 10-in.
    d. 12-in.

2. What type of rupture or breakage may occur when a pipe is unevenly supported along its length?

    a. Horizontal rupture
    b. Shear breakage
    c. Vertical rupture
    d. Beam breakage

3. Prestressed pipe is reinforced with

    a. Cement with fiberglass mesh
    b. Wire steel mesh
    c. Wire strands under tension
    d. Steel wrapping

4. What is AWWA's recommended maximum distance between valves in a residential area?

    a. 250 ft
    b. 500 ft
    c. 800 ft
    d. 1,000 ft

5. How much space should there be between the shoring and the sides of the excavation?

    a. None
    b. 1 in.
    c. 2 in.
    d. 3 in.

INSTALL UNITS

Sample Questions for Class III, answers on p. 207

1. What is the minimum diameter of a water main to be used for the installation of a fire hydrant?

    a. 2 in.
    b. 4 in.
    c. 6 in.
    d. 8 in.

2. What is the best location for a tap on a main?

    a. On top of the main
    b. On the side of the main
    c. 45° down from the top of the main
    d. 45° up from the bottom of the main

3. How many inches above the ground surface should the breakaway flange on fire hydrants be located?

    a. 2 in.
    b. 6 in.
    c. 10 in.
    d. 12 in.

4. Which of the following types of valves should be installed between fire hydrants and mains?

    a. Butterfly
    b. Auxiliary
    c. Foot
    d. Main

5. What is the best method to use to avoid contamination of a main when installing a new service connection?

    a. Shut valves on either side of the main, drill hole, thread hole, then install fitting
    b. Shut valves on either side of the main, drill hole, then weld saddle with fitting over the hole
    c. Leave main pressurized and install fitting by wet tap
    d. Leave main pressurized and install Rollins-Vorsky insertion tap

INSTALL UNITS

Sample Questions for Class IV, answers on p. 208

1. What backflow prevention device has a center relief valve?

    a. Double check valve assembly

    b. Atmospheric vacuum breaker

    c. Pressure vacuum breaker assembly

    d. Reduced pressure backflow assembly

2. As a general rule, water pipes should be separated from sewer pipes by a horizontal distance of

    a. 6 ft

    b. 8 ft

    c. 10 ft

    d. 12 ft

3. Fire hydrants should generally be set back from the curb by at least

    a. 1 ft

    b. 2 ft

    c. 3 ft

    d. 4 ft

4. How many valves should be installed on main intersections?

    a. 1

    b. 2

    c. 3

    d. 4

5. Which of the following is associated with using a pressure tap to make connections of new mains to an existing main?

    a. Discolored water

    b. Large amounts of water loss

    c. Low probability of contamination

    d. Loss of fire protection

OPERATE AND MAINTAIN EQUIPMENT

Operate Equipment	Class I	Class II	Class III	Class IV
Blowers and compressors	Application	Application	Application	Analysis
Cathodic protection devices	Comprehension	Comprehension	Application	Analysis
Centrifugal pumps	Application	Application	Analysis	Analysis
Chemical feeders	Analysis	Analysis	Analysis	Analysis
Chlorinators	Analysis	Analysis	Analysis	Analysis
Computers	Comprehension	Application	Application	Application
Electrical motors	Comprehension	Comprehension	Application	Application
Engines	Application	Application	Analysis	Analysis
Generators	Application	Application	Analysis	Analysis
Hydrants	Analysis	Analysis	Analysis	Analysis
Hydraulic equipment	Application	Application	Application	Application
Instrumentation	Comprehension	Comprehension	Comprehension	Comprehension
Leak correlators/detectors	Comprehension	Comprehension	Application	Analysis
Pipe and valve locators	Comprehension	Application	Application	Application
Positive-displacement pumps	Application	Application	Analysis	Analysis
Power tools	Application	Application	Application	Application
SCADA system/RTU/Telemetry/GIS	Comprehension	Comprehension	Application	Application
Tapping equipment	Application	Application	Application	Analysis
Valves	Analysis	Analysis	Analysis	Analysis
Perform Maintenance				
Backflow prevention devices	Comprehension	Comprehension	Comprehension	Comprehension
Blowers and compressors	Application	Application	Application	Analysis
Cathodic protection devices	Application	Application	Application	Application
Chemical feeders	Analysis	Analysis	Analysis	Analysis
Chlorinators	Analysis	Analysis	Analysis	Analysis
Corrosion control	Comprehension	Comprehension	Application	Application
Electric motors	Comprehension	Comprehension	Application	Application
Engines	Application	Application	Analysis	Analysis
Fittings	Application	Application	Analysis	Analysis
Generators	Application	Application	Analysis	Analysis
Hydrants	Analysis	Analysis	Analysis	Analysis
Hydraulic equipment	Application	Application	Application	Application
Instrumentation	Comprehension	Comprehension	Comprehension	Comprehension
Joints	Application	Application	Application	Application
Leak detection	Application	Application	Analysis	Analysis
Meters	Application	Application	Analysis	Analysis

Operate Equipment	Class I	Class II	Class III	Class IV
Pigs	Application	Application	Analysis	Analysis
Pipe repair	Application	Application	Analysis	Analysis
Pressure sensors	Application	Application	Application	Application
Pumps	Application	Application	Analysis	Analysis
Service connection	Application	Application	Analysis	Analysis
System flushing	Application	Application	Application	Application
Valves	Application	Application	Analysis	Analysis
Water storage facility	Application	Application	Analysis	Analysis

Suggestions for Study:

- Ability to adjust equipment
- Ability to calibrate equipment
- Ability to diagnose/troubleshoot equipment
- Ability to differentiate between preventive & corrective maintenance
- Ability to discriminate between normal and abnormal conditions
- Ability to evaluate operation of equipment
- Ability to follow written procedures
- Ability to record information
- Knowledge of corrosion control, dechlorination and disinfection processes
- Knowledge of data acquisition techniques
- Knowledge of different types of cross-connections and approved backflow methods and devices
- Knowledge of emergency plans
- Knowledge of facility operation and maintenance
- Knowledge of function of tools
- Knowledge of general electrical, mechanical and hydraulic principles
- Knowledge of internal combustion engines
- Knowledge of lubricant and fluid characteristics
- Knowledge of pipe fittings and joining methods
- Knowledge of piping material, type and size
- Knowledge of pneumatics
- Knowledge of protective coatings and paints
- Knowledge of regulations and standards
- Knowledge of safety procedures
- Knowledge of start-up and shut-down procedures

OPERATE AND MAINTAIN EQUIPMENT

Sample Questions for Class I, answers on p. 209

1. Which type of hydrant has no main valve but has a separate valve for each nozzle?
    a. Wet-barrel
    b. Warm-climate
    c. Dry-barrel
    d. Breakaway

2. What is the maximum theoretical suction lift of a centrifugal pump at sea level?
    a. 10 ft
    b. 34 ft
    c. 52 ft
    d. 85 ft

3. What type of motor is the simplest of all AC motors, with rotors consisting of a series of bars placed in slots?
    a. Squirrel-cage
    b. Wound-rotor
    c. Capacitor-start
    d. Synchronous

4. What does SCADA stand for?
    a. Statistical Calculations and Data Analysis
    b. Supervisory Control and Data Acquisition
    c. Standard Computer and Data Accessory
    d. Sample Concentration and Data Analyzer

5. Which of the following type of valve is used in maintaining prime to a pump?
    a. Foot
    b. Suction
    c. Vacuum header
    d. Butterfly

OPERATE AND MAINTAIN EQUIPMENT

Sample Questions for Class II, answers on p. 210

1. What type of cleaning plug has hardened steel or silicon carbide wire brushes?

    a. Scraping pig
    b. Bare pig
    c. Coated pig
    d. Drying pig

2. For best results, what is the minimum flushing velocity when using the unidirectional flushing method?

    a. 3 fps
    b. 5 fps
    c. 7 fps
    d. 10 fps

3. Calculate the cross-sectional area of a pipe that is 10 in. in diameter.

    a. 0.54 sq ft
    b. 4.00 sq ft
    c. 7.85 sq ft
    d. 8.33 sq ft

4. Which of the following characteristics of a pump is shown in a pump curve?

    a. Wire-to-water horsepower
    b. Motor horsepower
    c. Friction loss
    d. Efficiency

5. Which type of level sensor requires an air supply?

    a. Bubbler tube
    b. Diaphragm element
    c. Float mechanism
    d. Direct electronic sensor

OPERATE AND MAINTAIN EQUIPMENT

Sample Questions for Class III, answers on p. 211

1. What type of motor is used when infrequent starting is required and the load needs to be brought up to speed very quickly?

    a. Repulsion-induction
    b. Capacitor-start
    c. Wound-rotor
    d. Synchronous

2. Which type of level sensor has contained air connected to a pressure transducer via a tube?

    a. Bubbler tube
    b. Direct electronic sensor
    c. Float mechanism
    d. Diaphragm element

3. What type of instrumentation should be used to monitor distant locations?

    a. Telemetry systems
    b. Satellite systems
    c. Relay systems
    d. Progressive scan systems

4. What is the first action an operator should take if a large main break occurs?

    a. Call the fire department
    b. Locate the valves that isolate the leak
    c. Call the police department
    d. Notify the media

5. A meter indicates water is flowing from a fire hydrant at 2.6 cfm. How many gallons will flow from the hydrant if it is flushed for 35 minutes?

    a. 20 gal
    b. 91 gal
    c. 683 gal
    d. 910 gal

OPERATE AND MAINTAIN EQUIPMENT

Sample Questions for Class IV, answers on p. 212

1. Which of the following valves should be used when it becomes necessary to supply water at different pressure zones?

    a. Relief

    b. Rotary

    c. Pressure-reducing

    d. Needle

2. Which of the following pumps should be used to increase pressure at large distribution systems?

    a. Positive displacement

    b. Progressive cavity

    c. Vertical turbine

    d. Airlift

3. What should the flow meter read in gallons per minute, if a 12-in. diameter main is to be flushed at 5.0 fps?

    a. 1,764 gpm

    b. 2,900 gpm

    c. 3,600 gpm

    d. 3,920 gpm

4. AWWA recommends ⅝-in. meters be tested every

    a. 5 years

    b. 8 years

    c. 10 years

    d. 20 years

5. What is the term for electrochemical corrosion caused by the joining of two different metals?

    a. Concentration cell corrosion

    b. Bimetallic corrosion

    c. Tuberculation corrosion

    d. Pitting corrosion

SAFETY

Perform Work in a Safe Manner	Class I	Class II	Class III	Class IV
Chemical handling	Application	Application	Analysis	Analysis
Confined space entry	Application	Application	Analysis	Analysis
Contamination	Application	Application	Analysis	Analysis
Electrical grounding	Comprehension	Comprehension	Comprehension	Comprehension
Excavation	Application	Application	Analysis	Analysis
Fire safety	Application	Application	Application	Application
Lock-out/tag-out	Application	Application	Analysis	Analysis
Personal protective equipment	Application	Application	Analysis	Analysis
Traffic/work zone	Application	Application	Analysis	Analysis
Trenching and shoring	Application	Application	Analysis	Analysis

Suggestions for Study:

- Ability to communicate verbally and in writing
- Ability to demonstrate safe work habits
- Ability to follow written procedures
- Ability to interpret Material Safety Data Sheets
- Ability to recognize unsafe work conditions or potential safety hazards
- Ability to select and operate safety equipment
- Knowledge of emergency plans
- Knowledge of safety procedures
- Knowledge of regulations

SAFETY

Sample Questions for Class I, answers on p. 213

1. Depending on its concentration and length of exposure, which of the following gases can cause lung and skin irritation?
    a. Carbon monoxide
    b. Chlorine
    c. Methane
    d. Nitrogen

2. What class of fire involves electrical equipment?
    a. Class A
    b. Class B
    c. Class C
    d. Class D

3. When a permit is required to enter a confined space, who may sign the permit?
    a. Entrant
    b. Person attending
    c. Entry supervisor
    d. OSHA representative

4. When any piece of electrical equipment is being worked on, the circuit breaker should be
    a. Painted when repair is complete
    b. Videotaped for future reference
    c. De-energized and locked out
    d. Replaced

5. OSHA is the acronym for
    a. Organization for Safe Health Administration
    b. Occupational Safety and Health Administration
    c. Occupational, Safety and Health Act
    d. Organization of State Health Administrator

SAFETY

Sample Questions for Class II, answers on p. 214

1. What is the primary health risk of trihalomethanes?

    a. Cancer

    b. Liver damage

    c. Pancreas disorders

    d. Nervous system damage

2. What information is required on the tag during a switch lock-out?

    a. Social security number

    b. Name of person who locked out the switch

    c. Temperature of area

    d. Exact time the tag will be removed

3. Which of the following provides a profile of hazardous substances or mixtures?

    a. CERCLA

    b. OSHA

    c. CFR

    d. MSDS

4. What type of personal protective equipment should an operator wear when handling hypochlorites?

    a. Reflective vest

    b. Mask

    c. Ear plugs

    d. Eye goggles

5. Permit-required confined space entry requires

    a. Bright orange jacket, rubber boots, and gloves

    b. Chest or full body harness and a retrieval line

    c. Tool belt with flashlight attached

    d. Utility belt with a full complement of tools

SAFETY

Sample Questions for Class III, answers on p. 215

1. An atmosphere is defined as oxygen deficient if it contains less than what percent oxygen by volume?

    a. 19.5%

    b. 19.8%

    c. 20.5%

    d. 21.0%

2. During a confined space entry, how often must the confined space be monitored for hazardous atmospheres?

    a. Continuously

    b. Every 5 minutes

    c. Before entry only

    d. Before entry and then once per hour during entry

3. Under any soil conditions, cave-in protection is required for trenches or excavations that are how many feet deep?

    a. 2 ft

    b. 3 ft

    c. 4 ft

    d. 5 ft

4. Which of the following is colorless, odorless, lighter than air, highly flammable, and sometimes called swamp gas?

    a. Hydrogen sulfide

    b. Methane

    c. Carbon dioxide

    d. Radon

5. Material safety data sheets (MSDSs) are required for

    a. All chemicals used in the workplace regardless of hazard

    b. Only chemicals with known health hazards

    c. Only flammable or explosive chemicals

    d. Only chemicals with suspected health hazards

SAFETY

Sample Questions for Class IV, answers on p. 216

1. Fire protection facilities for each community are evaluated by
    a. USEPA
    b. ISO
    c. SDWA
    d. City or county government

2. Continued inhalation of radon gas is considered to contribute to which of the following?
    a. Gastrointestinal disease
    b. Hepatitis
    c. Cholera
    d. Lung cancer

3. When work is being performed in streets, the flagman should be positioned at least how many feet in front of the work space?
    a. 5 ft
    b. 20 ft
    c. 100 ft
    d. 500 ft

4. Which health effect category refers to an organic chemical that is a known carcinogen?
    a. Category I
    b. Category II
    c. Category III
    d. Category IV

5. All occupied trenches 4 or more ft deep must provide exits at
    a. 15-ft intervals
    b. 20-ft intervals
    c. 25-ft intervals
    d. 30-ft intervals

PERFORM ADMINISTRATIVE DUTIES

Perform Administrative Duties	Class I	Class II	Class III	Class IV
Write reports (internal, state, federal)	Application	Application	Analysis	Analysis
Promote customer service program	Comprehension	Comprehension	Analysis	Analysis
Promote media relations program	Comprehension	Comprehension	Application	Analysis
Promote public information program	Comprehension	Comprehension	Application	Analysis
Respond to complaints	Analysis	Analysis	Analysis	Analysis
Administer safety program	Application	Application	Analysis	Analysis
Administer compliance program	Application	Application	Analysis	Analysis
Administer emergency preparedness program	Application	Application	Analysis	Analysis
Develop budget	Comprehension	Application	Application	Analysis
Develop operation and maintenance plan	Application	Application	Analysis	Analysis
Plan and organize work activities	Application	Application	Analysis	Analysis
Part 141 National Primary Drinking Water Regulations				
Subpart A— General definitions	Comprehension	Comprehension	Comprehension	Comprehension
Subpart B— Maximum contaminant levels	Comprehension	Comprehension	Comprehension	Comprehension
Subpart C—Monitoring and analytical requirements	Comprehension	Comprehension	Comprehension	Comprehension
Subpart D—Reporting and recordkeeping	Comprehension	Comprehension	Comprehension	Comprehension
Subpart I—Control of lead and copper	Comprehension	Comprehension	Comprehension	Comprehension
Subpart Q—Public notification of drinking water violations	Comprehension	Comprehension	Comprehension	Comprehension

Suggestions for Study:

- Ability to assess likelihood of disaster occurring
- Ability to communicate verbally and in writing
- Ability to coordinate emergency response with other organizations
- Ability to determine what information needs to be recorded
- Ability to evaluate facility performance
- Ability to interpret and transcribe data
- Ability to organize information and review reports
- Ability to perform basic math
- Ability to perform impact assessments
- Ability to recognize unsafe work conditions and potential safety hazards
- Ability to translate technical language into common terminology
- Knowledge of facility operation and maintenance
- Knowledge of monitoring and reporting requirements
- Knowledge of operation and maintenance practices
- Knowledge of principles of finance
- Knowledge of principles of management and supervision
- Knowledge of principles of public relations
- Knowledge of public notification requirements
- Knowledge of public participation process
- Knowledge of recordkeeping function, policies, and requirements
- Knowledge of regulations

PERFORM ADMINISTRATIVE DUTIES

Sample Questions for Class I, answers on p. 217

1. The SDWA defines a public water system that supplies piped water for human consumption as one that has

    a. 10 service connection or serves 20 or more people for 60 or more days per year

    b. 15 service connections or serves 20 or more people for 90 or more days per year

    c. 10 service connections or serves 25 or more people for 30 or more days per year

    d. 15 service connections or serves 25 or more people for 60 or more days per year

2. The most common water complaints are taste, odor, and

    a. pH level

    b. Faucet pressure

    c. Colored water

    d. Leaking pipes

3. What US agency establishes drinking water standards?

    a. AWWA

    b. USEPA

    c. NIOSH

    d. NSF

4. Why should the operator contact area companies with underground utilities before starting an underground repair job?

    a. To determine if there have been recent excavations in that location

    b. To ask the companies to locate and mark the location of their utilities in the area of the repair job

    c. To determine if they also have excavating to do in the area

    d. To ask if they will help route traffic while you are doing the repair job

5. According to the USEPA regulations, the owner or operator of a public water system that fails to comply with applicable monitoring requirements must give notice to the public within

    a. 1 week of the violation in a letter hand-delivered to customers

    b. 45 days of the violation by posting a notice at the town hall

    c. 3 months of the violation in a daily newspaper in the area served by the system

    d. 1 year of the violation by including a letter with the water bill

PERFORM ADMINISTRATIVE DUTIES

Sample Questions for Class II, answers on p. 218

1. Your department uses 80 units of an item per week. You are required to maintain a 10-week reserve of this item at all times and it requires 4 weeks to obtain a new supply. What is the minimum reorder point?

    a. 320 units
    b. 800 units
    c. 1,120 units
    d. 2,240 units

2. What is the maximum contaminant level goal for chloride?

    a. 2.5 mg/L
    b. 25 mg/L
    c. 250 mg/L
    d. 2,500 mg/L

3. What is the median value of the following data: 100, 300, 580, 250, 275, 335, 580

    a. 250
    b. 300
    c. 346
    d. 580

4. The National Primary Drinking Water Regulations apply to drinking water contaminants that may have adverse effects on

    a. Water color
    b. Water taste
    c. Water odor
    d. Human health

5. Which of the following is considered an acute risk to health?

    a. Two Tier 2 violations
    b. One Tier 2 violation
    c. Two Tier 1 violations
    d. One Tier 1 violation

PERFORM ADMINISTRATIVE DUTIES

Sample Questions for Class III, answers on p. 219

1. Records on turbidity analyses should be kept for a minimum of
   a. 5 years
   b. 7 years
   c. 10 years
   d. 25 years

2. What is the difference between a primary standard and a secondary standard?
   a. Primary standards refer to substances that are carcinogenic, secondary standards do not
   b. Primary standards refer to substances that are thought to pose a threat to human health, secondary standards do not
   c. Primary standards refer to substances that, if not put in check, will eventually kill humans, secondary standards do not
   d. Secondary qualities are aesthetic qualities and will only make some people sick, while primary standards refer to substances that will make everyone sick and may possibly cause death

3. An employee receives an hourly wage of $13.25 plus overtime pay of 1.5 times the hourly wage. Overtime pay is given for each hour worked over 40 hours per week. If an employee works 48 hours during a week, what is the compensation before taxes?
   a. $159.00
   b. $530.00
   c. $689.00
   d. $848.00

4. What type of computer software is recommended for maintaining records such as turbidity levels?
   a. Word processor
   b. E-mail
   c. Graphics
   d. Database

5. How should a supervisor handle a recurring problem with an operator?

    a. Document the problem in writing and talk to the operator

    b. Promote the operator so he develops a better work ethic

    c. Ask a co-worker to discuss the problem with the operator

    d. Ignore the situation; problems tend to work themselves out

PERFORM ADMINISTRATIVE DUTIES

Sample Questions for Class IV, answers on p. 220

1. Records on bacteriological analyses should be kept for a minimum of
   a. 5 years
   b. 7 years
   c. 10 years
   d. 25 years

2. What should a supervisor do if an employee is performing work in an unsafe manner?
   a. Discuss the incident with the employee during the next performance appraisal
   b. Stop the work immediately and train the employee to perform the work safely
   c. Call OSHA immediately to investigate the incident
   d. Give the employee a written warning that the work was performed unsafely

3. Your plant's current annual budget for salaries is $450,000. A 3% raise will be given to all employees next year. Calculate the salary budget for next year.
   a. $463,500
   b. $470,000
   c. $483,500
   d. $490,000

4. What is the action level for lead?
   a. 0.01 mg/L
   b. 0.015 mg/L
   c. 0.1 mg/L
   d. 0.5 mg/L

5. How must water systems serving a population over 10,000 distribute the consumer confidence report to customers?
   a. Television announcement
   b. Radio announcement
   c. Public meeting
   d. Mail

ADDITIONAL SAMPLE QUESTIONS, answers on p. 221

1. A 6-in. pipeline needs to be flushed. If the desired length of pipeline to be flushed is 316 ft, how many minutes will it take to flush the line at 31 gpm?

    a. 10 minutes
    b. 15 minutes
    c. 30 minutes
    d. 60 minutes

2. What is the area of a trench that is 22.4 ft long and 3.3 ft wide?

    a. 26 sq ft
    b. 74 sq ft
    c. 143 sq ft
    d. 187 sq ft

3. Which of the following is typically associated with trihalomethanes?

    a. High levels of carbon dioxide in a surface water source
    b. Surface water high in inorganics
    c. Water with organics that has been chlorinated
    d. Groundwater or surface water high in inorganics

4. Which of the following mandates the language and methods for public notification?

    a. American Water Works Association
    b. National Rural Water Association
    c. US Environmental Protection Agency
    d. California State University

5. What is the primary source of lead in drinking water?

    a. Lakes
    b. Rivers near lead mines
    c. Corrosion of plumbing systems
    d. Groundwater

6. Interior copper tubing is usually joined by
   - a. Solder ✓
   - b. Flare
   - c. Compression
   - d. Union

7. Fire demand can account for what percent of the total capacity of a storage system?
   - a. 20%
   - b. 30%
   - c. 40%
   - d. 50% ✓

8. Which type of piping is most resistant to corrosion?
   - a. Plastic ✓
   - b. Stainless steel
   - c. Concrete
   - d. Asbestos-cement

9. Which of the following metals makes the best anode?
   - a. Brass
   - b. Cast iron
   - c. Zinc ✓
   - d. Copper

10. What is the minimum contact time when using the chlorine tablet method of disinfection?
    - a. 6 hours
    - b. 12 hours
    - c. 18 hours
    - d. 24 hours ✓

11. Piping containing reclaimed water should be painted what color?

    a. Orange
    b. Yellow
    c. Purple ✓
    d. Red

12. What is the pounds per square inch pressure at the bottom of a tank, if the water level is 38.29 ft?

    a. 7.3 psi
    b. 16.6 psi ✓
    c. 53.9 psi
    d. 88.4 psi

13. If a trench is 644 ft long, 3.5 ft wide, and 5.0 ft deep, approximately how many cubic yards of soil were excavated?

    a. 140 cu yd
    b. 300 cu yd
    c. 420 cu yd ✓
    d. 450 cu yd

14. Calculate the average pH given the following data.

1	2	3	4	5	6	7	8
7.45	7.49	7.56	7.63	7.60	7.54	7.52	7.41

    a. 7.4
    b. 7.5 ✓
    c. 30.1
    d. 60.2

15. What percent is 34,411 of 74,818?

    a. 34.411%
    b. 45.993% ✓
    c. 74.818%
    d. 217.42%

16. A solution was found to be 63.5% calcium hypochlorite. What is the milligrams per liter of calcium hypochlorite in the solution?

    a. 1,000
    b. 10,000
    c. 63,500
    d. 635,000

17. What is the diameter of a tank with a circumference of 408.2 ft?

    a. 31 ft
    b. 130 ft
    c. 260 ft
    d. 1,282 ft

18. What is the difference between a weak acid and a strong acid?

    a. Amount of hydrogen ions released
    b. Amount of buffering released
    c. Amount of hydroxyl ions released
    d. Amount of carbonate ions released

19. Which of the following analytical tests is the most important?

    a. Coliform bacteria
    b. Iron
    c. Manganese
    d. Hardness

20. Fuel oils, gasoline, and other organic compounds may permeate which type of piping?

    a. Plastic
    b. Asbestos-cement
    c. Fiberglass
    d. Concrete

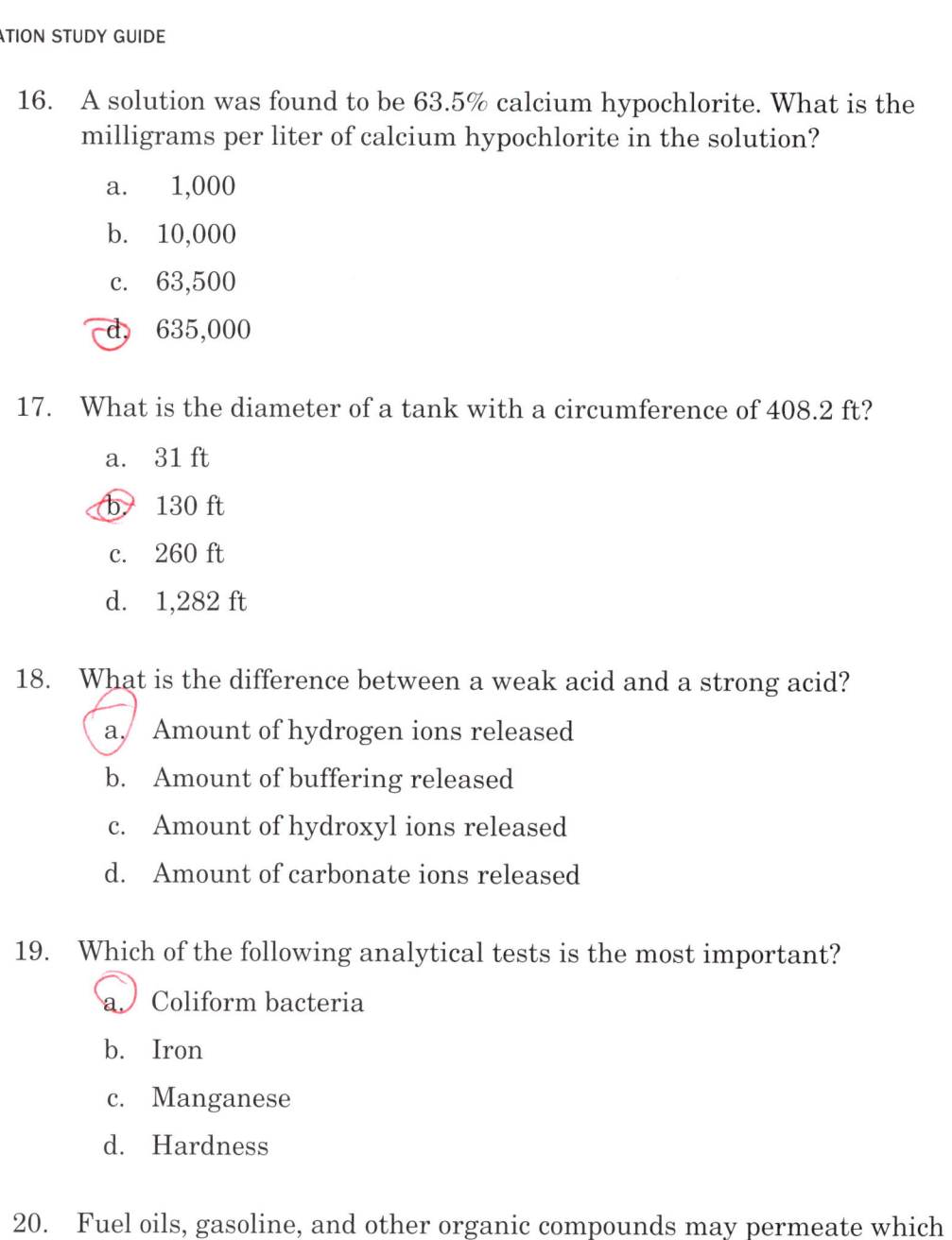

21. How long should a new tank be disinfected before a water sample is collected for coliform analysis?

    a. 6 hours

    b. 12 hours

    c. 24 hours

    d. 48 hours

22. What type of concrete reservoir is constructed much like a swimming pool, is rather shallow, and difficult to cover?

    a. Cast-in-place concrete

    b. Hydraulically applied concrete-lined

    c. Circular prestressed-concrete

    d. Prestressed concrete-wire-wound

23. Which of the following types of valve is the best to use to maintain the water level in a tank?

    a. Altitude

    b. Tapping

    c. Butterfly

    d. Needle

24. Why should isolation valves be installed at frequent intervals in distribution piping?

    a. To stop water from flowing backward through a pump that is not in operation

    b. To throttle flow and maintain a lower pressure in the lower distribution system zone

    c. So that air that accumulates at high points in pipes can be automatically vented

    d. So that small sections of water main may be shut off for maintenance

25. What is the internal surface area of a cylindrical tank (bottom, top, and the cylinder wall), if it is 125.0 ft in diameter and 48.5 ft high?

    a. 6,063 sq ft

    b. 19,036 sq ft

    c. 24,531 sq ft

    d. 43,567 sq ft

26. A trench that averages 4.2 ft wide and 5.4 ft in depth is dug for the purpose of installing a 24-in. diameter pipeline. If the trench is 1,287 ft long, how much soil in cubic feet will be put in the trench after the pipe is in place, assuming that the only soil left over is that which the pipe now occupies?

    a. 1,300 cu ft
    b. 4,000 cu ft
    c. 25,000 cu ft
    d. 29,000 cu ft

27. Customers usually notice turbidity readings in excess of how many nephelometric turbidity units?

    a. 2
    b. 3
    c. 4
    d. 5

28. What is the recommended chlorine dosage when disinfecting a water main using the tablet method?

    a. 25 mg/L
    b. 50 mg/L
    c. 75 mg/L
    d. 100 mg/L

29. What water quality problem is most likely to occur at dead-end water mains?

    a. Dirty water
    b. Taste and odor
    c. Milky water due to air bubbles
    d. Dirty clothes due to manganese

30. Asbestos-cement pipe contains which of the following material?

    a. Fiberglass
    b. Silica sand
    c. Polyethylene
    d. Silica lime

31. What is the primary purpose of well grouting?

    a. Protect the well casing from corrosion

    b. Seal off poor-quality aquifers

    c. Prevent surface water contamination from entering the water that is being drawn into the well

    d. Protect the casing by restraining unstable earth materials

32. Which type of butterfly valve connection is best to use underground?

    a. Mechanical joint

    b. Flanged joint

    c. Lug wafer joint

    d. Check joint

33. What type of concrete tank is made like a home basement but with more reinforcement?

    a. Cast-in-place concrete

    b. Hydraulically applied concrete-lined

    c. Circular prestressed-concrete

    d. Prestressed concrete-wire-wound

34. The pressure at the lowest bottom point of a clear well is 13.2 psi. What is the total water depth of the clear well at its lowest point?

    a. 23.1 ft

    b. 30.5 ft

    c. 61.0 ft

    d. 69.3 ft

35. How many hours must the water remain undisturbed in a pipe before collecting a sample for lead and copper analysis?

    a. 4 hours

    b. 6 hours

    c. 8 hours

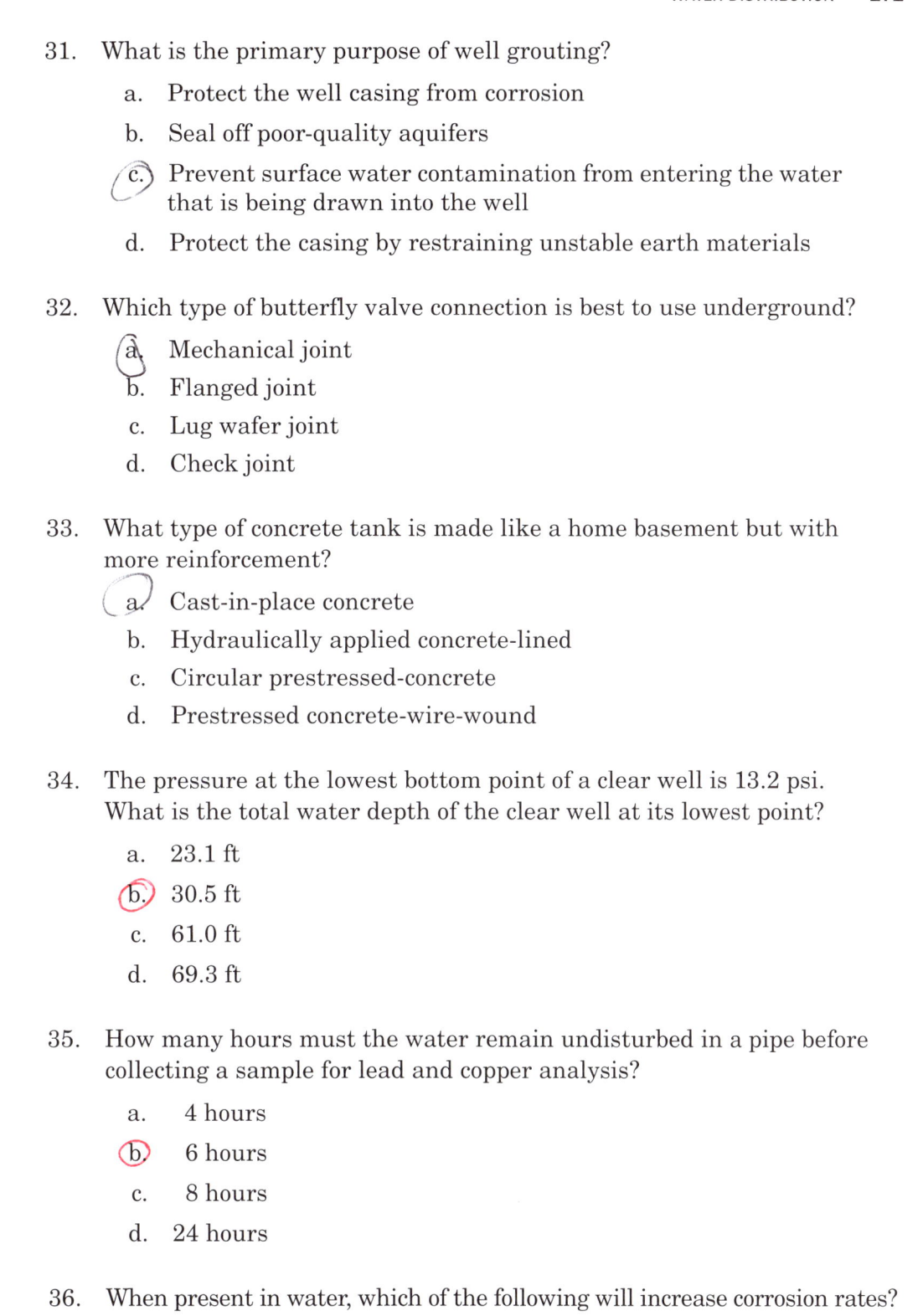

    d. 24 hours

36. When present in water, which of the following will increase corrosion rates?

    a. Zinc orthophosphate

    b. Dissolved oxygen

    c. Silicate compounds

    d. Bicarbonate ion

37. What is the recommended chlorine dosage when disinfecting a water main using the slug method?

    a. 50 mg/L
    b. 100 mg/L
    c. 125 mg/L
    d. 300 mg/L

38. If a confined aquifer's recharge area is elevated, the water in the aquifer will be

    a. Contaminated
    b. High in iron and manganese
    c. Under pressure
    d. Soft

39. Which one of the following service line materials may cause corrosion when connected to brass valves?

    a. Galvanized iron
    b. PVC
    c. Copper
    d. Plastic

40. When is the best time to conduct a leak survey?

    a. Night, when noise is lowest
    b. During low water usage
    c. When the ground has been dry awhile
    d. During periods of high water usage

41. PVC pipe can use tees, elbows, and other fittings from what other type of pipe?

    a. Prestressed concrete cylinder
    b. Ductile iron
    c. Concrete
    d. Asbestos-cement

42. Water use for peak hour demand is determined by the highest 1-hour period during one

    a. Day
    b. Week
    c. Month
    d. Year

43. The zone of influence is defined as the
    a. Distance from a well that the cone of depression affects the normal water level
    b. Depression around a well of the water surface caused by pumping water from the well
    c. Change in water elevation from the normal level to the pumping level
    d. Measured distance from the ground to the pumping level

44. The secondary standard for iron recommends that the concentration not exceed
    a. 0.30 mg/L
    b. 0.40 mg/L
    c. 0.50 mg/L
    d. 0.60 mg/L

45. What is the name of the pipe that serves as the inlet and outlet to an elevated tank?
    a. Pedestal
    b. Riser
    c. Main pedestal
    d. Service

46. What type of ductile-iron pipe joint is quickly assembled and is the least expensive?
    a. Flanged joint
    b. Mechanical joint
    c. Push-on joint
    d. Ball-and-socket joint

47. Where is the most common location of leaks on old service connections?
    a. Household plumbing
    b. Connections to the curb stop
    c. Connections to the gate valve
    d. Pressure regulator

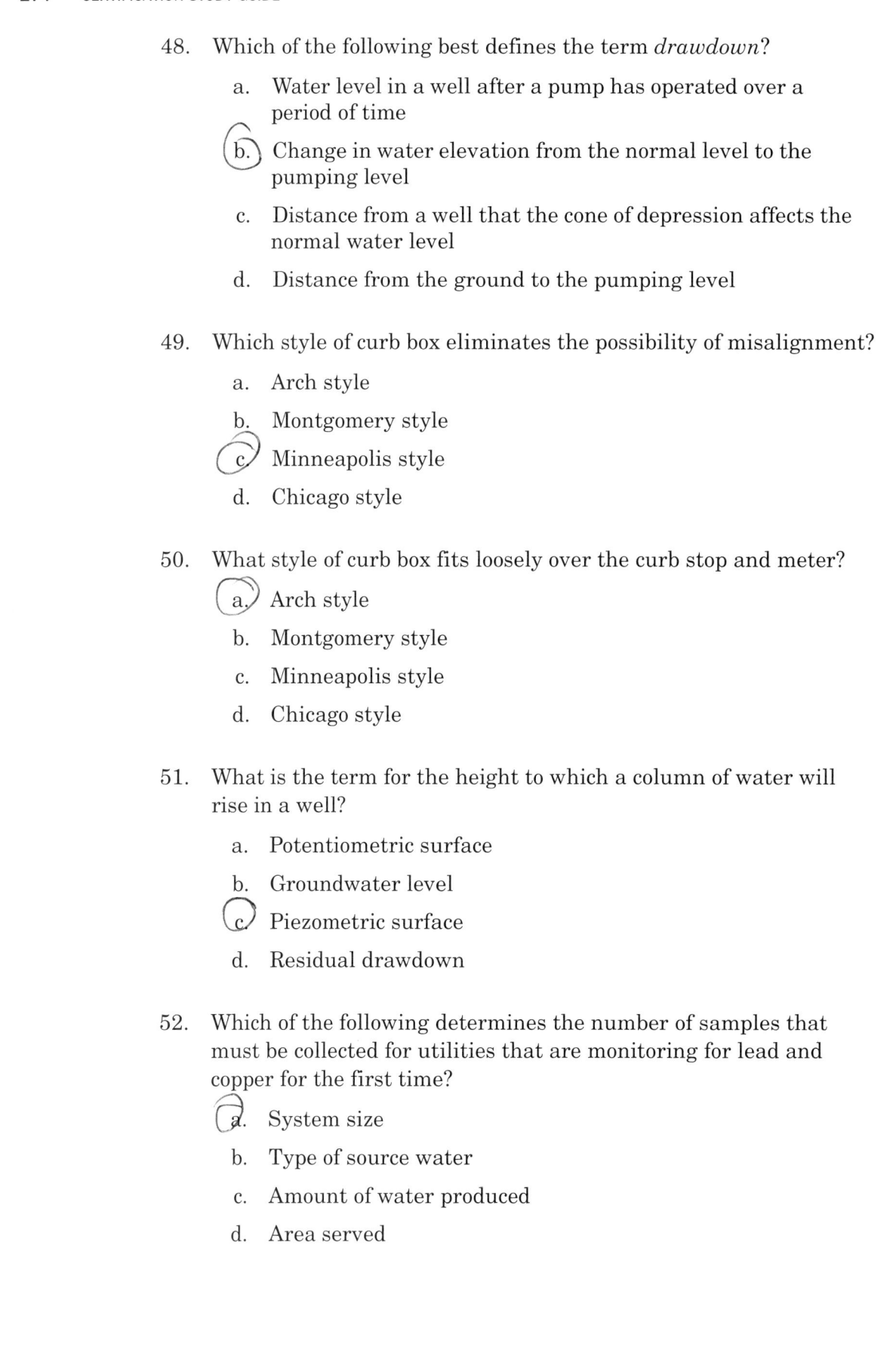

48. Which of the following best defines the term *drawdown*?

    a. Water level in a well after a pump has operated over a period of time

    (b.) Change in water elevation from the normal level to the pumping level

    c. Distance from a well that the cone of depression affects the normal water level

    d. Distance from the ground to the pumping level

49. Which style of curb box eliminates the possibility of misalignment?

    a. Arch style

    b. Montgomery style

    (c.) Minneapolis style

    d. Chicago style

50. What style of curb box fits loosely over the curb stop and meter?

    (a.) Arch style

    b. Montgomery style

    c. Minneapolis style

    d. Chicago style

51. What is the term for the height to which a column of water will rise in a well?

    a. Potentiometric surface

    b. Groundwater level

    (c.) Piezometric surface

    d. Residual drawdown

52. Which of the following determines the number of samples that must be collected for utilities that are monitoring for lead and copper for the first time?

    (a.) System size

    b. Type of source water

    c. Amount of water produced

    d. Area served

53. What type of level sensor cannot be used where the water body freezes?
    a. Bubbler tube
    b. Diaphragm element
    c. Float mechanism
    d. Direct electronic sensor

54. Current meters are also known as
    a. Venturi meters
    b. Insertion meters
    c. Orifice meters
    d. Velocity meters

55. An example of a large-flow water meter is
    a. Velocity type
    b. Positive displacement type
    c. Rotating disk type
    d. Piston type

56. What type of meter is primarily used for measuring dirty or corrosive water?
    a. Magnetic
    b. Insertion
    c. Ultrasonic
    d. Venturi

57. What type of meter determines the total flow by multiplying the flow through the meter by a factor?
    a. Proportional
    b. Magnetic
    c. Insertion
    d. Current

58. Which type of hydrant is hidden from public view?
    a. Breakaway
    b. Flush
    c. Wet-barrel
    d. Warm-climate

59. What type of pump is most often used by public water systems?

    a. Piston

    b. Velocity

    c. Positive-displacement

    d. Archimedes

60. Dry-barrel fire hydrants are used in what type of environment?

    a. Dry

    b. Wet

    c. Freezing

    d. Hot

61. In general, what is the maximum practical lift of pumps?

    a. 5 to 15 ft

    b. 15 to 25 ft

    c. 30 to 40 ft

    d. 45 to 55 ft

62. When using a power tool to open or close a distribution valve, it is best to

    a. Use low speed

    b. Do last few turns manually

    c. Do first few turns manually to loosen valve

    d. Use high speed

63. Which hydrant type has no main valve but each nozzle has a separate valve?

    a. Breakaway

    b. Dry barrel

    c. Wet-barrel

    d. Flush

64. What type of pump has rotating impellers within a pump case?

    a. Axial-flow

    b. Centrifugal

    c. Jet

    d. Mixed-flow

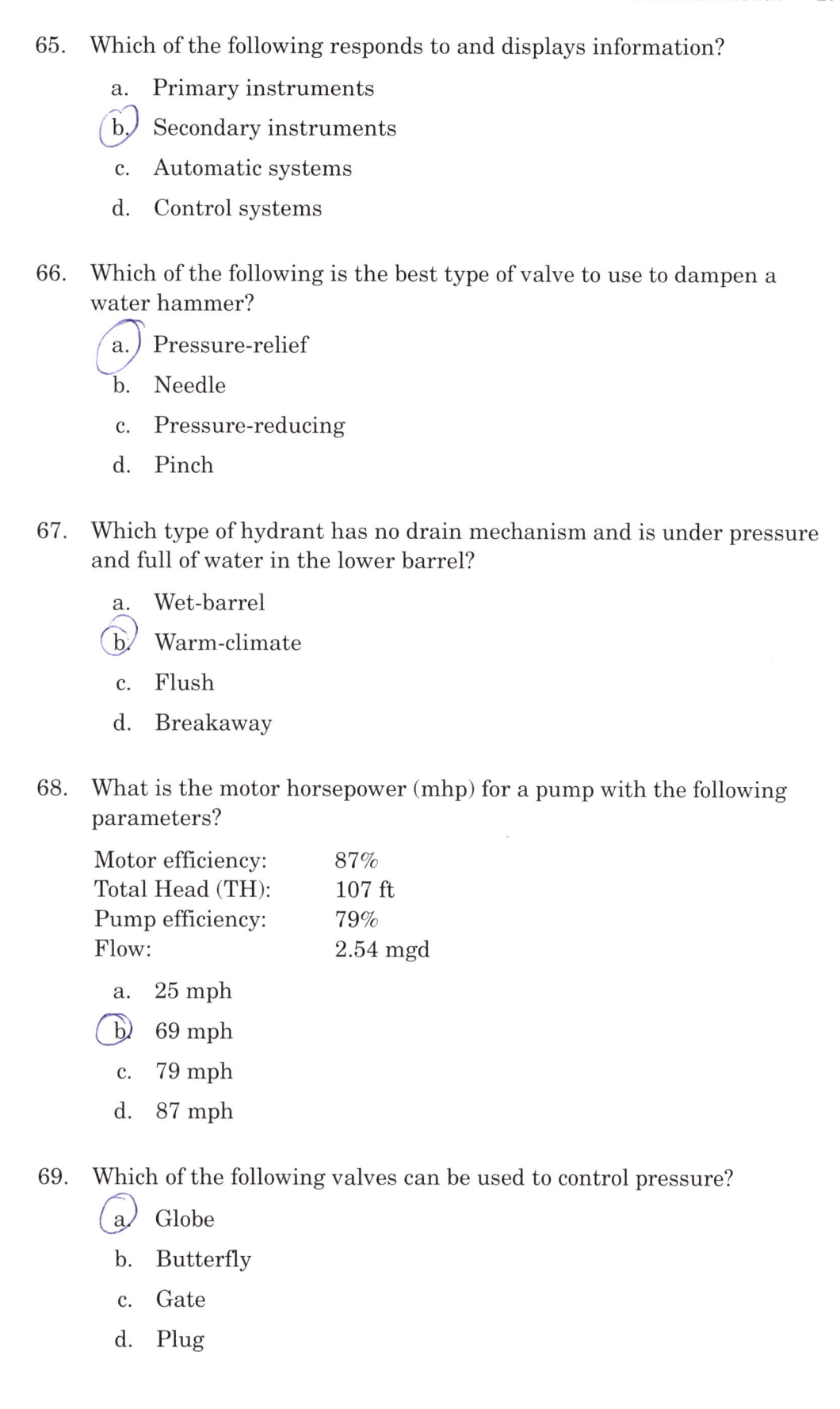

65. Which of the following responds to and displays information?

    a. Primary instruments

    b. Secondary instruments

    c. Automatic systems

    d. Control systems

66. Which of the following is the best type of valve to use to dampen a water hammer?

    a. Pressure-relief

    b. Needle

    c. Pressure-reducing

    d. Pinch

67. Which type of hydrant has no drain mechanism and is under pressure and full of water in the lower barrel?

    a. Wet-barrel

    b. Warm-climate

    c. Flush

    d. Breakaway

68. What is the motor horsepower (mhp) for a pump with the following parameters?

    Motor efficiency: 87%
    Total Head (TH):   107 ft
    Pump efficiency: 79%
    Flow: 2.54 mgd

    a. 25 mph

    b. 69 mph

    c. 79 mph

    d. 87 mph

69. Which of the following valves can be used to control pressure?

    a. Globe

    b. Butterfly

    c. Gate

    d. Plug

70. What type of motor requires no windings and has low starting torque?

    a. Split-phase

    b. Squirrel cage

    c. Repulsion-induction

    d. Capacitor-start

71. What type of pump has high efficiencies (90 to 95%) at very high pressure, impellers that are very close fitting, and high maintenance costs?

    a. Jet

    b. Vertical turbine

    c. Mixed flow

    d. Centrifugal

72. What type of motor has a starting current seldom over the full-load operating current, has a stator, and the resistance of the rotor circuit can be controlled when the motor is running?

    a. Synchronous

    b. Squirrel-cage

    c. Compound

    d. Wound-rotor

73. Which of the following valves is best to use in throttling flow situations?

    a. Globe

    b. Butterfly

    c. Gate

    d. Ball

74. Which type of temperature sensor uses two wires of different material?

    a. Thermistor

    b. Thermoresistor

    c. Thermocouple

    d. Thermo-electrocouple

75. What is the pressure head on a system exerting a static pressure of 62 psi?

   a. 27 ft

   b. 89 ft

   c. 143 ft

   d. 175 ft

76. What is the difference in temperature in degrees Fahrenheit between the water source and the processed water in a storage tank, if the temperature of the water source is 12.6°C and the water in the storage tank is 16.3°C?

   a. 3.7°F

   b. 6.7°F

   c. 35.7°F

   d. 38.7°F

77. What type of meter is based on the Doppler effect?

   a. Magnetic meter

   b. Insertion meter

   c. Ultrasonic meter

   d. Venturi meter

78. Which of the following is a measure of a material's opposition to the flow of electric current?

   a. Ohms

   b. Amperes

   c. Volts

   d. Watts

79. Which type of direct electronic sensor for level measurement is placed above the water level?

   a. Variable-resistance level sensor

   b. Probe level system

   c. Ultrasonic level-sensing system

   d. Diaphragm level-sensing system

80. Which of the following is most likely a cause of decreasing drawdown and an unchanging static water level?

    a. Aquifer is being pumped in excess of recharge

    b. Well screen is becoming worn

    c. Pump is losing efficiency

    d. Cone of depression has changed

81. What is the pressure head at a fire hydrant in feet, if the pressure gauge reads 121 psi?

    a. 52 ft

    b. 86 ft

    c. 141 ft

    d. 280 ft

82. Which of the following devices is used to measure water depths in storage facilities?

    a. Transducer

    b. Magnetic sensor

    c. Venturi meter

    d. Thermistor

83. What is the maximum recommended head loss for positive displacement water meters that are less than 2 in.?

    a. 5 psi

    b. 10 psi

    c. 15 psi

    d. 20 psi

84. Which of the following devices can check water flow in the distribution system?

    a. Pitot meter

    b. Amperometric meter

    c. Flowtech meter

    d. Service line meter

85. Which of the following best defines the term *stray-current corrosion*?

    a. Reaction between a metal and water

    b. Decomposition of a metal by its reaction with an acidic water

    c. Decomposition of a material caused by an outside electric current

    d. Reaction between two different metals with water acting as an electrolyte

86. What is the optimal maximum water velocity through pipes under normal operating conditions?

    a. 4 fps

    b. 5 fps

    c. 6 fps

    d. 7 fps

87. Which type of temperature sensor uses a semiconductive material?

    a. Thermocouple

    b. Thermoresistor

    c. Thermistor

    d. Thermo-electrocouple

88. AWWA recommends ⅝-in. meters be tested every

    a. 5 years

    b. 8 years

    c. 10 years

    d. 20 years

89. What class of fire involves oil or grease?

    a. Class A

    b. Class B

    c. Class C

    d. Class D

90. What health risk is associated with nitrate in water?

    a. Liver damage

    b. Methemoglobinemia

    c. Kidney damage

    d. Nervous system damage

91. What is the secondary maximum contaminant level for copper?
    a. 0.1 mg/L
    b. 0.5 mg/L
    c. 1.0 mg/L
    d. 2.0 mg/L

92. The secondary standard for manganese recommends that the concentration not exceed
    a. 0.02 mg/L
    b. 0.03 mg/L
    c. 0.04 mg/L
    d. 0.05 mg/L

93. Where are maximum contaminant levels for water quality indicators specified?
    a. Webster's New World Dictionary
    b. National Primary Drinking Water Regulations
    c. Fair Labor Standards Act
    d. Standard Methods for the Examination of Water and Wastewater

94. Why is it essential to ventilate a valve vault before entry?
    a. Remove excessive moisture
    b. Equalize temperature and pressure
    c. Eliminate foul odors
    d. Remove dangerous gases

95. What class of fire involves sodium or magnesium?
    a. Class A
    b. Class B
    c. Class C
    d. Class D

96. Which of the following parameters is used to indicate the clarity of water?

    a. pH

    b. Chlorine residual

    c. Turbidity

    d. Bacteriological

97. The primary purpose of heck valves is to prevent

    a. Excessive pump pressure

    b. Priming

    c. Water from flowing in two directions

    d. Water hammer

98. If only two rings of packing are used in the stuffing box, how many degrees should the joints be staggered?

    a. 15–45 degrees

    b. 45–90 degrees

    c. 90–180 degrees

    d. 180–225 degrees

99. What is pressure head caused by?

    a. Water flow

    b. Water pressure

    c. Gauge pressure

    d. Water elevation

100. What is the purpose of a bypass valve on a larger size gate valve?

    a. Connect a new main to an existing main

    b. Reduce pressure across both sides to ease opening and closing

    c. Increase flow through the main line

    d. Allow easy location of the main valve

184 CERTIFICATION STUDY GUIDE

MATH FOR MORE PRACTICE, answers on p. 233

1. Records on a pump indicate that it pumped 1,228,400 gal during the 31 days of May. Calculate the average gallons pumped per day.
    a. 12,284
    b. 39,626
    c. 122,840
    d. 390,626

2. Convert 81.67% to decimal form.
    a. 0.8167
    b. 8.167
    c. 81.67
    d. 816.7

3. Convert 0.118 to percent.
    a. 0.118%
    b. 1.18%
    c. 11.8 %
    d. 118%

4. What is the percent lime in a slurry, if 12 lb of lime are mixed in a 55-gal drum that contains 52 gal of water?
    a. 2.7%
    b. 4.5%
    c. 5.4%
    d. 9.0%

5. What percent is 15,325 of 21,611 to the nearest tenth of a percent?
    a. 6.3%
    b. 14.1%
    c. 36.9%
    d. 70.9%

6. Solve the proportion: 6:72 = x:480

    a. x = 40
    b. x = 72
    c. x = 408
    d. x = 414

7. Chlorine is fed at a rate of 25 lb/day for a flow rate of 7 cfs. To maintain the same dosage, what adjustment to the chlorinator should be made when the flow rate is increased to 12 cfs?

    a. 30 lb/day
    b. 43 lb/day
    c. 50 lb/day
    d. 63 lb/day

8. Convert 34 cfs to gallons per minute.

    a. 255 gpm
    b. 2,040 gpm
    c. 7,150 gpm
    d. 15,300 gpm

9. A solution is 5.8% hypochlorite. What is the milligrams per liter of hypochlorite in the solution?

    a. 58 mg/L
    b. 580 mg/L
    c. 5,800 mg/L
    d. 58,000 mg/L

10. What is the circumference of a tank that is 75.0 ft in diameter?

    a. 150 ft
    b. 236 ft
    c. 563 ft
    d. 626 ft

11. What is the pounds per square inch of pressure at the bottom of a tank, if the water level is 33.11 ft deep?

    a. 1.4 psi

    b. 3.1 psi

    c. 14.3 psi

    d. 76.5 psi

12. The pressure head at a fire hydrant is 195 ft. Calculate the pounds per square inch.

    a. 51.5 psi

    b. 84.4 psi

    c. 101 psi

    d. 195 psi

13. Find the area in square feet for a rectangular-shaped basin that is 672 ft in length and 68.5 ft in width.

    a. 215 sq ft

    b. 2,110 sq ft

    c. 1,895 sq ft

    d. 46,032 sq ft

14. Find the area of a triangle in square inches, if it has a height of 14 in. and a base of 17 in.

    a. 119 sq in.

    b. 196 sq in.

    c. 238 sq in.

    d. 289 sq in.

15. Calculate the volume of a trench that is 363 ft in length, 3.5 ft in width, and 6.0 ft in depth.

    a. 373 cu ft

    b. 2,723 cu ft

    c. 7,623 cu ft

    d. 9,529 cu ft

16. A trench is 430 ft long, 4.0 ft wide, and 5.5 ft deep. How many cubic yards of soil are excavated?

    a. 215 cu yd

    b. 350 cu yd

    c. 788 cu yd

    d. 9,460 cu yd

17. How many gallons are there in a pipe that is 18 in. in diameter and 216 ft long?

    a. 1,908 gal

    b. 2,246 gal

    c. 2,430 gal

    d. 2,861 gal

18. A tank is conical at the bottom and cylindrical at the top. The diameter of the cylinder is 21.0 ft with a depth of 42.0 ft and the cone depth is 14.2 ft. What is the volume of the tank in cubic feet?

    a. 1,639 cu ft

    b. 12,901 cu ft

    c. 14,540 cu ft

    d. 16,179 cu ft

19. Given the following data, calculate the average in pounds per day for chlorine used.

Mon.	Tue.	Wed.	Thur.	Fri.	Sat.	Sun.
34	28	26	29	31	32	35

    a. 26.0 lb/day

    b. 30.7 lb/day

    c. 107.5 lb/day

    d. 214.9 lb/day

20. What is the average million gallons per day flow from a storage tank given the following data? Note: All measured values were to the nearest 0.1 mgd.

Mon.	Tue.	Wed.	Thur.	Fri.	Sat.	Sun.
6.7	6.4	6.8	7.5	7.7	7.3	6.8

    a. 6.4 mgd

    b. 6.8 mgd

    c. 7.0 mgd

    d. 7.7 mgd

21. What is the velocity of flow in feet per second for a 10-in. diameter pipe, if it delivers 550 gpm?

    a. 0.8 fps

    b. 1.2 fps

    c. 1.8 fps

    d. 2.2 fps

22. Water is flowing in a pipeline at 2.15 cfs. What is the flow in gallons per minute?

    a. 16 gpm

    b. 21 gpm

    c. 161 gpm

    d. 968 gpm

23. A water tank with a capacity of 3.0 mil gal is being filled at a rate of 4,810 gpm. How many hours will it take to fill the tank?

    a. 1.4 hours

    b. 3.0 hours

    c. 10.4 hours

    d. 28.8 hours

24. The velocity through a channel is 2.10 fps. If the channel is 6.5 ft wide and 2.8 ft in water depth, what is the flow in cubic feet per second?

    a. 6 cfs

    b. 14 cfs

    c. 28 cfs

    d. 38 cfs

WATER DISTRIBUTION   189

25. A water tank has a volume of 2.0 mil gal and the flow from the tank is 5.36 mgd. What is the detention time in hours?

   a. 8.96 hr

   b. 12.34 hr

   c. 23.56 hr

   d. 53.60 hr

26. Given the following data, determine the detention time in hours.

   - Distribution pipe from water plant to storage tank is 2,485 ft in length and 12-in. in diameter
   - Storage tank averages 1,875,000 gal of water at any given time
   - Flow through system is 6.82 mgd

   a. 6.65 hr

   b. 12.55 hr

   c. 14.63 hr

   d. 18.89 hr

27. An 8-in. main line needs to be flushed. The length of pipeline to be flushed is 250 ft. How many minutes will it take to flush the line at 25 gpm?

   a. 7 minutes

   b. 13 minutes

   c. 26 minutes

   d. 31 minutes

28. A water treatment plant injects chlorine at a dosage rate of 3.00 mg/L after the filters. The chlorine residual is 1.45 mg/L at a distant point in the distribution system. Calculate the chlorine demand between the two sampling points.

   a. 1.45 mg/L

   b. 1.55 mg/L

   c. 2.90 mg/L

   d. 4.45 mg/L

29. What is the chlorine residual in a system that has a chlorine dosage of 2.75 mg/L and a chlorine demand of 1.93 mg/L.

    a. 0.82 mg/L

    b. 1.75 mg/L

    c. 4.67 mg/L

    d. 5.31 mg/L

30. How many pounds per day of chlorine gas are required to treat 8.65 mgd, if the dosage is 2.75 mg/L?

    a. 11 lb/day

    b. 24 lb/day

    c. 72 lb/day

    d. 198 lb/day

31. How many pounds per day of chlorine are needed to treat 38.75 mgd, if the dosage is 3.50 mg/L?

    a. 42 lb/day

    b. 136 lb/day

    c. 323 lb/day

    d. 1,131 lb/day

32. What should the setting be on a chlorinator in pounds per day, if the dosage desired is 2.90 mg/L and the pumping rate from the well is 975 gpm?

    a. 29 lb/day

    b. 34 lb/day

    c. 41 lb/day

    d. 336 lb/day

33. A small tank containing 1,500 gal of water needs to be disinfected in order to be put back in service. If the dosage needed is 35 mg/L, how many pounds of calcium hypochlorite (60.5% available chlorine) are required?

    a. 0.73 lb

    b. 1.5 lb

    c. 25 lb

    d. 43 lb

34. How many gallons of 6.00% sodium hypochlorite solution are needed to disinfect a 1.0-ft diameter pipeline that is 752 ft long. The required dosage is 25.0 mg/L.

    a. 0.92 gal
    b. 1.83 gal
    c. 4.43 gal
    d. 15.30 gal

35. A 2.50-mil gal storage tank needs to be disinfected with a 62.5% calcium hypochlorite solution. If the chlorine dosage desired is 50.0 mg/L, how many gallons of calcium hypochlorite solution are required?

    a. 125 gal
    b. 200 gal
    c. 1,040 gal
    d. 1,563 gal

36. Determine the number of gallons of 12.5% sodium hypochlorite solution needed to disinfect a water main that is 24.0 in. in diameter and the pipeline is 750 ft long, if the dosage required is 30.0 mg/L?

    a. 4.1 gal
    b. 14.1 gal
    c. 27.3 gal
    d. 36.0 gal

37. A 2.0-ft diameter pipe that is 1.4 miles long is disinfected with chlorine; 44 lb of chlorine are used. Calculate the dosage in milligrams per liter.

    a. 1.7 mg/L
    b. 3.1 mg/L
    c. 17 mg/L
    d. 31 mg/L

38. An operator mixes 125 lb of HTH (64.50% available chlorine) with 250 gal of water. What percent is the calcium hypochlorite solution?

    a. 4%
    b. 6%
    c. 8%
    d. 10%

39. A well produces 162 gpm. The drawdown for the well is 16 ft. Calculate the specific yield in gallons per minute per foot.

    a. 2 gpm/ft

    b. 5 gpm/ft

    c. 10 gpm/ft

    d. 16 gpm/ft

40. Find the drawdown of a well, if the well yields 265 gpm and the specific yield is 11.7 gpm/ft.

    a. 10.3 ft

    b. 11.7 ft

    c. 17.6 ft

    d. 22.7 ft

41. The static water level (nonpumping well) in a well is 84.5 ft. The pumping level is 104.2 ft. What is the drawdown?

    a. 19.7 ft

    b. 36.6 ft

    c. 45.1 ft

    d. 188.7 ft

42. The static level in the well is 79.12 ft and the drawdown is 26.08 ft. Calculate the pumping water level in the well.

    a. 11.3 ft

    b. 34.3 ft

    c. 53.0 ft

    d. 105.2 ft

43. A pump discharges 680 gpm. How many gallons will it discharge in 8 hours?

    a. 5,440 gal

    b. 130,560 gal

    c. 326,400 gal

    d. 408,000 gal

44. A hypochlorite solution is being pumped from a small tank that is 2.0 ft in diameter. The level in the tank drops 2.25 ft in 4 hours. How many gallons per minute of hypochlorite solution were used?

   a. 0.22 gpm

   b. 0.53 gpm

   c. 2.2 gpm

   d. 5.3 gpm

45. Find the total head, in feet, for a pump with a total static head of 19 ft and a head loss of 3.7 ft.

   a. 5.1 ft

   b. 15.3 ft

   c. 22.7 ft

   d. 70.3 ft

46. A motor with 89% efficiency is supplied with 25 hp. Calculate the brake horsepower.

   a. 3 bhp

   b. 22 bhp

   c. 25 bhp

   d. 28 bhp

47. Two columns of water are filled completely at sea level to a height of 88 ft. Column A is 0.5 in. in diameter. Column B is 5 in. in diameter. What will the two pressure gauges, each attached to the bottom of each column, read?

	Column A	Column B
a.	3.8 psi	38.0 psi
b.	8.8 psi	8.0 psi
c.	20.3 psi	20.3 psi
d.	38.0 psi	38.0 psi

48. Given the following data, calculate the total kilowatts needed to operate the following small facility when everything is running.

Raw water pump	=	300 hp
5 Flocculators	=	60 hp
Filter pump for backwashing	=	100 hp
Chlorination	=	25 hp
Clear water pump	=	100 hp
Lighting	=	11 hp
Instrumentation	=	4 hp

    a. 260 kW

    b. 448 kW

    c. 600 kW

    d. 1,386 kW

49. Given the following data for lead concentrations find the 90th percentile.

    0.010 mg/L   0.017 mg/L   0.009 mg/L   0.005 mg/L   0.006 mg/L
    0.019 mg/L   0.011 mg/L   0.010 mg/L   0.008 mg/L   0.013 mg/L

    a. 0.006 mg/L

    b. 0.010 mg/L

    c. 0.017 mg/L

    d. 0.019 mg/L

50. How many pounds of high test hypochlorite (HTH) are needed to make 500 gal of a 10.0% HTH solution?

    a. 50 lb

    b. 243 lb

    c. 375 lb

    d. 417 lb

Water Distribution

Answers

SYSTEM DESIGN

Sample Questions for Class I – Answers

1. Answer: **d.** Fire protection requirements

   Reference: *AWWA, Water Distribution Operator Training Handbook*, Second Edition, Page 36.

2. Answer: **b.** C value. The higher the C value of the pipe, the smoother the pipe. The Hazen-Williams formula is used to calculate pipe size.

   Reference: *AWWA, Water Distribution Operator Training Handbook*, Second Edition, Page 39.

3. Answer: **c.** Decreases the life of water heaters and other water-using appliances

   Reference: *AWWA, Water Distribution Operator Training Handbook*, Second Edition, Page 37.

4. Answer: **c.** Late at night to lessen traffic disruption and minimize customer complaints

   Reference: *AWWA, Water Distribution Operator Training Handbook*, Second Edition, Page 109.

5. Answer: **a.** 14.3 psi

   Solution: $\text{psi} = \dfrac{\text{depth, ft}}{2.31 \text{ ft/psi}} = \dfrac{33.11 \text{ ft}}{2.31 \text{ ft/psi}} = 13.3 \text{ psi}$

   Reference: *AWWA, Basic Science Concepts and Applications*, Page 249, and *AWWA, Water Distribution Operator Training Handbook*, Second Edition, Page 248.

SYSTEM DESIGN

Sample Class II Questions – Answers

1. Answer: **d.** Ball-and-socket

   Reference: *AWWA, Water Distribution Operator Training Handbook*, Second Edition, Page 45.

2. Answer: **a.** 1 to 2 ft

   Reference: *AWWA, Water Distribution Operator Training Handbook*, Second Edition, Page 74.

3. Answer: **b.** Grid

   Reference: *AWWA, Water Distribution Operator Training Handbook*, Second Edition, Page 33.

4. Answer: **b.** 0.54 sq ft

   Solution: area, sq ft = $(0.785)(D)^2$

   First, convert pipe diameter from inches to feet: 10 in./12 in. per foot = 0.83.

   area, sq ft = (0.785) (0.83 ft) (0.83 ft) = 0.54 sq ft

5. Answer: **b.** Total water use for a year by 365 days

   Reference: *AWWA, Water Distribution Operator Training Handbook*, Second Edition, Page 36.

SYSTEM DESIGN

Sample Class III Questions – Answers

1. Answer: **b.** 3 fps

   Solution: First, convert the number of gallons per minute to cubic feet per second.

   $$\text{no. of cfs} = \frac{722 \text{ gpm}}{(7.5 \text{ gal/cu ft})(60 \text{ sec/min})} = 1.61 \text{ cfs}$$

   Next, convert the diameter from inches to feet.
   no. of ft = (10 in.) (1 ft/12 in.) = 0.83 ft
   flow, cfs = (area, sq ft) (velocity, fps), where the area = $(0.785)(D)^2$
   1.6 cfs = (0.785) (0.83 ft) (0.83 ft) (velocity, fps)
   Rearrange the problem to solve for velocity:

   $$\text{velocity, fps} = \frac{1.6 \text{ cfs}}{(0.785)(0.83 \text{ ft})(0.83 \text{ ft})} = 2.96 \text{ fps, round to 3 fps}$$

   Reference: AWWA, *Principles and Practices of Water Supply Operations, Basic Science Concepts and Applications,* Second Edition, Page 322.

2. Answer: **b.** Tree

   Reference: AWWA, *Water Distribution Operator Training Handbook,* Second Edition, Page 33.

3. Answer: **a.** Contamination of the system by backsiphonage

   Reference: AWWA, *Water Distribution Operator Training Handbook,* Second Edition, Page 29 and 155.

4. Answer: **b.** Backpressure will not exist

   Reference: AWWA, *Water Distribution Operator Training Handbook,* Second Edition, Page 161.

5. Answer: **a.** 20 psi

   Reference: AWWA, *Water Distribution Operator Training Handbook,* Second Edition, Page 37.

SYSTEM DESIGN

Sample Questions for Class IV – Answers

1. Answer: **c.** 2,700,000 gal

   Solution: First, find the volume in gallons for each of the distribution pipes.
   volume, gal = (0.785) (D)^2 (length, ft) (7.5 gal/cu ft)
   volume, pipe "A" gal = (0.785) (3.0 ft) (3.0 ft) (1,376 ft) (7.5 gal/cu ft) = 72,910.8 gal
   volume, pipe "B" gal = (0.785) (2.0 ft) (2.0 ft) (833 ft) (7.5 gal/cu ft) = 19,617.15 gal

   Next, determine the volume of the storage tank.
   Equation is: volume, gal = (0.785) (D)^2 (height, ft) (7.5 gal/cu ft)
   Volume, gal = (0.785) (120 ft) (120 ft) (30.73 ft) (7.5 gal/cu ft) = 2,605,289.4 gal

   Last, add the three volumes together for the total volume.
   total volume, gal = 72,910.8 gal + 19,617.15 gal + 2,605,289.4 gal = 2,697,817.35 gal, rounded to 2,700,000

   Reference: AWWA, *Principles and Practices of Water Supply Operations, Basic Science Concepts and Applications,* Second Edition, Page 87, Example 3.

2. Answer: **b.** Low soil moisture content

   Reference: AWWA, *Water Distribution Operator Training Handbook*, Second Edition, Page 123.

3. Answer: **b.** Mechanical joint

   Reference: AWWA, *Principles and Practices of Water Supply Operations, Water Transmission and Distribution,* Second Edition, Page 35.

4. Answer: **d.** 4

   Reference: AWWA, *Water Distribution Operator Training Handbook*, Second Edition, Page 33.

5. Answer: **d.** 1,764

   Solution:
   First, convert the diameter of the pipe from inches to feet:
   no. of feet = (12 in.) (1 ft/12 in.) = 1 ft
   Next, calculate the pipe's cross-sectional area in square feet.
   area, sq ft = (0.785) (D)^2
   area, sq ft = (0.785) (1 ft) (1 ft) = 0.785 sq ft
   Next, find the flow in the pipe in cfs: flow, cfs = (area, sq ft) (velocity, fps)
   flow, cfs = (0.785 sq ft) (5 fps) = 3.92 cfs
   Last, determine the reading on the flow meter in gallons per minute.
   flow, gpm = (3.92 cfs) (7.5 gal/cu ft) (60 sec/min) = 1,764 gpm

MONITOR WATER QUALITY

Sample Questions for Class I – Answers

1. Answer: **a.** More effective

    Reference: AWWA, *Principles and Practices of Water Supply Operations, Water Treatment,* Second Edition, Page 175. *California State University, Small Water System Operation and Maintenance,* Third Edition, Chapter 4, Section 4.5.

2. Answer: **b.** 10 mg/L

    Reference: AWWA, *Principles and Practices of Water Supply Operations, Water Transmission and Distribution,* Second Edition, Page 206.

3. Answer: **b.** Increase the well's productivity

    Reference: AWWA, *Principles and Practices of Water Supply Operations, Water Sources*, Second Edition, Page 53.

4. Answer: **d.** Fecal material from warm-blooded animals

    Reference: AWWA, *Water Distribution Operator Training Handbook*, Second Edition, Page 23.

5. Answer: **c.** Hydrogen ions

    Reference: AWWA, *Principles and Practices of Water Supply Operations, Basic Science Concepts and Applications,* Second Edition, Chapter Chemistry 5, Page 426.

MONITOR WATER QUALITY

Sample Questions for Class II – Answers

1. Answer: **a.** 3 hours

   Reference: AWWA, *Water Distribution Operator Training Handbook*, Second Edition, Page 94.

2. Answer: **b.** Water level in a well when the well is not in operation. It is measured from the ground surface to the water surface.

   Reference: AWWA, *Water Distribution Operator Training Handbook*, Second Edition, Page 184.

3. Answer: **a.** At points representative of conditions within the system

   Reference: AWWA, *Water Distribution Operator Training Handbook*, Second Edition, Page 25, and AWWA, *Principles and Practices of Water Supply Operations, Water Quality,* Second Edition, Page 42.

4. Answer: **b.** Depression around the well of the water surface caused by pumping water from the well

   Reference: AWWA, *Water Distribution Operator Training Handbook*, Second Edition, Page 185.

5. Answer: **b.** Suspended particles

   Reference: AWWA, *Water Distribution Operator Training Handbook*, Second Edition, Page 28.

MONITOR WATER QUALITY

Sample Class III Questions – Answers

1. Answer: **c.** Time composite samples

   Reference: AWWA, *Water Distribution Operator Training Handbook*, Second Edition, Page 25.

2. Answer: **b.** Sodium thiosulfate

   Reference: AWWA, *Water Distribution Operator Training Handbook*, Second Edition, Page 27.

3. Answer: **b.** Bacteria

   Reference: AWWA, *Principles and Practices of Water Supply Operations, Water Quality,* Second Edition, Page 115, and AWWA, *Water Distribution Operator Training Handbook*, Second Edition, Page 27.

4. Answer: **c.** Water level below the normal level that persists after a well pump has been off for a period of time

   Reference: AWWA, *Water Distribution Operator Training Handbook*, Second Edition, Page 186.

5. Answer: **b.** Population

   Reference: Title 40 CFR 141.86 (c) and 141.86 (d) (4) (ii)

MONITOR WATER QUALITY

Sample Class IV Questions – Answers

1. Answer: **b.** At locations that are representative of conditions within the system

   Reference: *AWWA, Water Distribution Operator Training Handbook*, Second Edition, Page 25.

2. Answer: **a.** Temperature

   Reference: *AWWA, Principles and Practices of Water Supply Operations, Water Quality,* Second Edition, Chapter 6, Page 154.

3. Answer: **d.** 24 and 48 hours at 35°C

   Reference: *AWWA, Principles and Practices of Water Supply Operations, Water Quality,* Second Edition, Chapter 4, Page 110.

4. Answer: **d.** High

   Reference: *AWWA, Principles and Practices of Water Supply Operations, Water Transmission and Distribution,* Second Edition, Page 317, Table 11-1

5. Answer: **a.** 16.2

   Solution: specific yield, gpm/ft = $\dfrac{\text{well yield, gpm}}{\text{drawdown, ft}}$

   specific yield, gpm/ft = $\dfrac{365 \text{ gpm}}{22.5 \text{ ft}}$ 16.222 gpm/ft, round to 16.2 gpm/ft

INSTALL UNITS

Sample Class I Questions – Answers

1. Answer: **d.** Shoring

    Reference: *AWWA, Water Distribution Operator Training Handbook,* Second Edition, Page 78.

2. Answer: **b.** Galvanized iron

    Reference: *AWWA, Water Distribution Operator Training Handbook,* Second Edition, Page 129

3. Answer: **b.** Steel

    Reference: *AWWA, Water Distribution Operator Training Handbook,* Second Edition, Page 46, and *AWWA, Principles and Practices of Water Supply Operations, Water Transmission and Distribution,* Second Edition, Page 40.

4. Answer: **c.** Vent air that has accumulated in the well column while the well is not in use

    Reference: *AWWA, Principles and Practices of Water Supply Operations, Water Transmission and Distribution,* Second Edition, Page 60.

5. Answer: **b.** Filled with a solution of 25 ppm to 50 ppm free chlorine for at least 24 hours prior to flushing

    Reference: *AWWA, Principles and Practices of Water Supply Operations, Water Transmission and Distribution,* Second Edition, Chapter 5.

INSTALL UNITS

Sample Class II Questions – Answers

1. Answer: **a.** 6-in. Taps greater than 1-in. should be made through tapping saddles on water mains that are 6-in. or less in diameter.

   Reference: AWWA, *Water Distribution Operator Training Handbook*, Second Edition, Page 134.

2. Answer: **d.** Beam breakage

   Reference: AWWA, *Water Distribution Operator Training Handbook*, Second Edition, Page 42, and AWWA, *Principles and Practices of Water Supply Operations, Water Transmission and Distribution,* Second Edition, Page 18.

3. Answer: **c.** Wire strands under tension

   Reference: AWWA, *Principles and Practices of Water Supply Operations, Water Transmission and Distribution,* Second Edition, Page 49.

4. Answer: **c.** 800 ft

   Reference: AWWA, *Water Distribution Operator Training Handbook*, Second Edition, Page 34.

5. Answer: **a.** None

   Reference: AWWA, *Principles and Practices of Water Supply Operations, Water Transmission and Distribution,* Second Edition, Page 49.

INSTALL UNITS

Sample Questions for Class III – Answers

1. Answer: **c.** 6 in.

   Reference: *AWWA, Water Distribution Operator Training Handbook*, Second Edition, Page 36–37.

2. Answer: **c.** 45° down from the top of the main. A tap located directly on the top of a main is more likely to draw air into the service and a tap near the bottom could draw in sediment.

   Reference: *AWWA, Water Distribution Operator Training Handbook*, Second Edition, Page 134, and *AWWA, Principles and Practices of Water Supply Operations, Water Transmission and Distribution*, Second Edition, Chapter 9, page 269.

3. Answer: **a.** 2 in.

   Reference: *AWWA, Water Distribution Operator Training Handbook*, Second Edition, Page 102.

4. Answer: **b.** Auxiliary

   Reference: *AWWA, Water Distribution Operator Training Handbook*, Second Edition, Page 100.

5. Answer: **c.** Leave main pressurized and install fitting by wet tap

   Reference: *AWWA, Water Distribution Operator Training Handbook*, Second Edition, Page 132.

INSTALL UNITS

Sample Questions for Class IV – Answers

1. Answer: **d.** Reduced pressure backflow assembly

    Reference: AWWA, *Water Distribution Operator Training Handbook*, Second Edition, Page 158–159.

2. Answer: **c.** 10 ft

    Reference: AWWA, *Water Distribution Operator Training Handbook*, Second Edition, Page 72.

3. Answer: **b.** 2 ft

    Reference: AWWA, *Water Distribution Operator Training Handbook*, Second Edition, Page 102, and AWWA, *Principles and Practices of Water Supply Operations, Water Transmission and Distribution,* Second Edition, Page 167.

4. Answer: **c.** 3

5. Answer: **c.** Low probability of contamination

    Reference: AWWA, *Principles and Practices of Water Supply Operations, Water Transmission and Distribution,* Second Edition, Page 117.

OPERATE AND MAINTAIN EQUIPMENT

Sample Questions for Class I – Answers

1. Answer: **a.** Wet-barrel

   Reference: *AWWA, Water Distribution Operator Training Handbook*, Second Edition, Page 100.

2. Answer: **b.** 34 ft

   Reference: *AWWA, Water Distribution Operator Training Handbook*, Second Edition, Page 194.

3. Answer: **a.** Squirrel-cage

   Reference: *AWWA, Water Distribution Operator Training Handbook*, Second Edition, Page 204.

4. Answer: **b.** Supervisory Control and Data Acquisition

   Reference: *AWWA, Water Distribution Operator Training Handbook*, Second Edition, Page 230.

5. Answer: **a.** Foot

   Reference: *AWWA, Water Distribution Operator Training Handbook*, Second Edition, Page 199.

OPERATE AND MAINTAIN EQUIPMENT

Sample Questions for Class II – Answers

1. Answer: **a.** Scraping pig

   Reference: *AWWA, Water Distribution Operator Training Handbook*, Second Edition, Page 112.

2. Answer: **b.** 5 fps

   Reference: *AWWA, Water Distribution Operator Training Handbook*, Second Edition, Page 109.

3. Answer: **a.** 0.54 sq ft

   Solution: First, convert pipe diameter from inches to feet.
   10 in./12 in. per foot = 0.83 ft
   area, sq ft = (0.785) (D)^2
   area, sq ft = (0.785) (0.83 ft) (0.83 ft) = 0.54 sq ft

   Reference: *AWWA, Principles and Practices of Water Supply Operations, Basic Science Concepts and Applications,* Second Edition, Page 79.

4. Answer: **d.** Efficiency. Capacity, head, and required power are also shown in a pump curve.

   Reference: *AWWA, Principles and Practices of Water Supply Operations, Water Transmission and Distribution,* Second Edition, Page 554, Figure D-5.

5. Answer: **a.** Bubbler tube

   Reference: *AWWA, Water Distribution Operator Training Handbook*, Second Edition, Page 221.

OPERATE AND MAINTAIN EQUIPMENT

Sample Questions for Class III – Answers

1. Answer: **b.** Capacitor-start

    Reference: *AWWA, Water Distribution Operator Training Handbook,* Second Edition, Page 203.

2. Answer: **d.** Diaphragm element

    Reference: *AWWA, Water Distribution Operator Training Handbook,* Second Edition, Page 221.

3. Answer: **a.** Telemetry systems

    Reference: *AWWA, Water Distribution Operator Training Handbook,* Second Edition, Page 226.

4. Answer: **b.** Locate the valves that isolate the leak

    Reference: *AWWA, Water Distribution Operator Training Handbook,* Second Edition, Page 120.

5. Answer: **c.** 683 gal

    Solution: First, convert the flow in cubic feet per minute to gallons per minute.

    gpm = (2.6 cfm) (7.5 gal/cu ft) = 19.5 gpm

    Then determine the number of gallons that flowed through the fire hydrant.

    gallons = (19.5 gpm) (35 minutes) = 682.5 gal, rounded to 683

    Reference: *AWWA, Water Distribution Operator Training Handbook,* Second Edition, Page 4 and 247.

OPERATE AND MAINTAIN EQUIPMENT

Sample Questions for Class IV – Answers

1. Answer: **c.** Pressure-reducing

    Reference: *AWWA, Water Distribution Operator Training Handbook*, Second Edition, Page 62.

2. Answer: **c.** Vertical turbine

    Reference: *AWWA, Principles and Practices of Water Supply Operations, Water Transmission and Distribution,* Second Edition, Page 366.

3. Answer: **a.** 1,764 gpm

    Solution: First convert the diameter of the pipe from inches to feet.
    no. of feet = (12 in.) (1 ft/12 in.) = 1 ft
    Next, calculate the pipe's cross-sectional area in square feet.
    area, sq ft = (0.785) (D)^2
    area, sq ft = (0.785) (1 ft) (1 ft) = 0.785 sq ft
    Next, find the flow in the pipe in cubic feet per second.
    flow, cfs = (area, sq ft) (velocity, fps)
    flow, cfs = (0.785 sq ft) (5 fps) = 3.92 cfs
    Last, determine the reading on the flow meter in gallons per minute.
    flow, gpm = (3.92 cfs) (7.5 gal/cu ft) (60 sec/min) = 1,764 gpm

    Reference: *AWWA, Principles and Practices of Water Supply Operations, Basic Science Concepts and Applications,* Second Edition, Page 79, 320, and 323.

4. Answer: **c.** 10 years

    Reference: *AWWA, Water Distribution Operator Training Handbook*, Second Edition, Page 153.

5. Answer: **b.** Bimetallic corrosion

    Reference: *AWWA, Water Distribution Operator Training Handbook*, Second Edition, Page 124.

SAFETY

Sample Questions for Class I – Answers

1. Answer: **b.** Chlorine

   Reference: *AWWA, Principles and Practices of Water Supply Operations, Water Transmission and Distribution,* Second Edition, Chapter 5, Page 149.

2. Answer: **c.** Class C

   Reference: Title 29 CFR, Part 155 (c) (10).

3. Answer: **c.** Entry supervisor

   Reference: Title 29 CFR Part 1910.146(e) (2).

4. Answer: **c.** De-energized and locked out

   Reference: *AWWA, Principles and Practices of Water Supply Operations, Water Transmission and Distribution,* Second Edition, Chapter 13, Page 418.

5. Answer: **b.** Occupational Safety and Health Administration

   Reference: *AWWA, Principles and Practices of Water Supply Operations, Water Transmission and Distribution,* Second Edition, Chapter 11, Page 348.

SAFETY

Sample Questions for Class II – Answers

1. Answer: **a.** Cancer

    Reference: *AWWA, Water Distribution Operator Training Handbook*, Second Edition, Page 19–20.

2. Answer: **b.** Name of person who locked out the switch

    Reference: *AWWA, Principles and Practices of Water Supply Operations, Water Transmission and Distribution,* Second Edition, Chapter 12, Page 394.

3. Answer: **d.** MSDS

    Reference: *California State University, Very Small Water System Operation and Maintenance*, Third Edition, Chapter 6, Section 6.18.

4. Answer: **d.** Eye goggles

    Reference: *AWWA, Water Distribution Operator Training Handbook*, Second Edition, Page 217.

5. Answer: **b.** Chest or full body harness and a retrieval line

    Reference: Title 29 CFR Part 1910.146 (k) (3) (i).

SAFETY

Sample Questions for Class III – Answers

1. Answer: **a.** 19.5%

    Reference: Title 29 CFR Part 1910.146(b) Definitions.

2. Answer: **a.** Continuously

    Reference: Title 29 CFR Part 1910.146(c) (5) (ii) (E) and Appendix C, Example 1.

3. Answer: **d.** 5 ft

    Reference: AWWA, *Principles and Practices of Water Supply Operations, Water Transmission and Distribution,* Second Edition, Page 147.

4. Answer: **b.** Methane

    Reference: AWWA, *Principles and Practices of Water Supply Operations, Water Treatment,* Second Edition, Page 496, and the *California State University, Very Small Water System Operation and Maintenance,* Third Edition, Chapter 6, Section 6.18.

5. Answer: **a.** All chemicals used in the workplace regardless of hazard

    Reference: Title 29 CFR Part 1910.1200.

SAFETY

Sample Questions for Class IV – Answers

1. Answer: **b.** ISO. The Insurance Services Office, Inc., rates communities using the ISO Public Protection Classification Program.

    Reference: AWWA, *Water Distribution Operator Training Handbook*, Second Edition, Page 36.

2. Answer: **d.** Lung cancer

    Reference: AWWA, *Water Distribution Operator Training Handbook*, Second Edition, Page 24.

3. Answer: **c.** 100 ft

    Reference: *California State University, Very Small Water System Operation and Maintenance,* Third Edition, Chapter 6, Section 6.52.

4. Answer: **a.** Category I

    Reference: AWWA, *Principles and Practices of Water Supply Operations, Water Quality,* Second Edition, Page 171.

5. Answer: **c.** 25-ft intervals

    Reference: Title 29 CFR, Part 1926, Subpart P.

PERFORM ADMINISTRATIVE DUTIES

Sample Questions for Class I – Answers

1. Answer: **d.** 15 service connections or serves 25 or more people for 60 or more days per year

    Reference: AWWA, *Water Distribution Operator Training Handbook*, Second Edition, Page 16.

2. Answer: **c.** Colored water

    Reference: AWWA, *Water Distribution Operator Training Handbook*, Second Edition, Page 210.

3. Answer: **b.** USEPA

    Reference: AWWA, *Water Distribution Operator Training Handbook*, Second Edition, Page 15.

4. Answer: **b.** To ask the companies to locate and mark the location of their utilities in the area of the repair job

    Reference: AWWA, *Principles and Practices of Water Supply Operations, Water Transmission and Distribution,* Second Edition, Page 90.

5. Answer: **c.** 3 months of the violation in a daily newspaper in the area served by the system

    Reference: Title 40 CFR 141.32(b)(1).

PERFORM ADMINISTRATIVE DUTIES

Sample Questions for Class II – Answers

1. Answer: **c.** 1,120 units

   Solution: (10 weeks) (80 units/week) = 800 units reserve required
   (4-week order time) (80 units/week) = 320 units order time
   800 + 320 = 1,120 units

2. Answer: **c.** 250 mg/L

   Reference: AWWA, *Principles and Practices of Water Supply Operations, Water Quality,* Second Edition, Chapter 1, Table 1-3, Page 13.

3. Answer: **b.** 300

   Solution: Arrange data in ascending order: 100, 250, 275, 300, 335, 580, 580; then find the middle value.

4. Answer: **d.** Human health

   Reference: AWWA, *Water Distribution Operator Training Handbook,* Second Edition, Page 17.

5. Answer: **c.** Two Tier 1 violations

   Reference: Title 40 CFR, Part 141.

PERFORM ADMINISTRATIVE DUTIES

Sample Questions for Class III – Answers

1. Answer: **a.** 5 years

    Reference: *AWWA, Water Distribution Operator Training Handbook,* Second Edition, Page 19.

2. Answer: **b.** Primary standards refer to substances that are thought to pose a threat to human health, secondary standards do not.

    Reference: *AWWA, Water Distribution Operator Training Handbook,* Second Edition, Page 17.

3. Answer: **c.** $689.00

    Solution: (40 hours) ($13.25/hour) + (8 hours) ($13.25/hour) (1.5 overtime pay)
    = $530.00 + $159.00 overtime = $689.00

4. Answer: **d.** Database

    Reference: *AWWA, Principles and Practices of Water Supply Operations, Water Transmission and Distribution,* Second Edition, Page 469.

5. Answer: **a.** Document the problem in writing and talk to the operator

PERFORM ADMINISTRATIVE DUTIES

Sample Questions for Class IV – Answers

1. Answer: **a.** 5 years

    Reference: *AWWA, Water Distribution Operator Training Handbook*, Second Edition, Page 19.

2. Answer: **b.** Stop the work immediately and train the employee to perform the work safely. The supervisor should also document the incident in the daily log, not as a formal written reprimand or warning, in the event of future similar events.

    Reference: No reference source specified

3. Answer: **a.** $463,500

    Solution: current salary budget + 3% raise = $450,000 + 3% ($450,00)
    = $450,000 + ($450,000 × 0.03) = $450,000 + $13,500 = $463,500

    Reference: *AWWA, Water Distribution Operator Training Handbook*, Second Edition, Page 2.

4. Answer: **b.** 0.015 mg/L

    Reference: *AWWA, Principles and Practices of Water Supply Operations, Water Quality,* Second Edition, Page 24.

5. Answer: **d.** Mail

    Reference: *AWWA, Water Distribution Operator Training Handbook*, Second Edition, Page 20.

WATER DISTRIBUTION 221

ADDITIONAL SAMPLE QUESTIONS, Answers

1. Answer: **b.** 15 minutes

    Solution: First, convert the pipe diameter from inches to feet.

    no. of feet = (6 in.) (1 ft/12 in.) = 0.50 ft

    Next, determine the volume in gallons for 316 ft of the 6-in. pipeline.

    Formula: volume, gal = (0.785) (D)^2 (length, ft) (7.5 gal/cu ft)

    volume, gal = (0.785) (0.50 ft) (0.50 ft) (316 ft) (7.5 gal/cu ft) = 465 gal

    Then, find the flushing time.

    $$\text{flushing time, min} = \frac{465 \text{ gal}}{31 \text{ gpm}} = 15 \text{ min}$$

    Reference: AWWA, *Principles and Practices of Water Supply Operations, Basic Science Concepts and Applications,* Second Edition, Page 85 and 200.

2. Answer: **b.** 74 sq ft

    Solution:

    area = (length) (width)

    area = (22.4 ft) (3.3 ft) = 73.92 sq ft, round to 74

    Reference: AWWA, *Principles and Practices of Water Supply Operations, Basic Science Concepts and Applications,* Second Edition, Page 74.

3. Answer: **c.** Water with organics that has been chlorinated

    Reference: AWWA, *Water Distribution Operator Training Handbook,* Second Edition, Page 19.

4. Answer: **c.** US Environmental Protection Agency

    Reference: AWWA, *Water Distribution Operator Training Handbook,* Second Edition, Page 17.

5. Answer: **c.** Corrosion of plumbing systems

    Reference: AWWA, *Water Distribution Operator Training Handbook,* Second Edition, Page 19 and 208.

6. Answer: **a.** Solder

    Reference: AWWA, *Principles and Practices of Water Supply Operations, Water Transmission and Distribution,* Second Edition, Page 258.

7. Answer: **d.** 50%

   Reference: AWWA, *Principles and Practices of Water Supply Operations, Water Transmission and Distribution,* Second Edition, Page 181 and AWWA, *Water Distribution Operator Training Handbook,* Second Edition, Page 166.

8. Answer: **a.** Plastic pipe

   Reference: AWWA, *Water Distribution Operator Training Handbook,* Second Edition, Page 39 and 43.

9. Answer: **c.** Zinc

   Reference: AWWA, *Water Distribution Operator Training Handbook,* Second Edition, Page 124.

10. Answer: **d.** 24 hours

    Reference: AWWA, *Water Distribution Operator Training Handbook,* Second Edition, Page 93.

11. Answer: **c.** Purple

    Reference: No reference source specified

12. Answer: **b.** 16.6 psi

    Solution: $\text{psi} = \dfrac{\text{depth, ft}}{2.31 \text{ ft/psi}}$

    $\text{psi} = \dfrac{38.29 \text{ ft}}{2.31 \text{ ft/psi}} = 16.6 \text{ psi}$

    Reference: AWWA, *Water Distribution Operator Training Handbook,* Second Edition, Page 248.

13. Answer: **c.** 420 cu yd

    Solution: First, find the number of cubic feet.
    volume, cu ft = (length, ft) (width, ft) (depth, ft)
    volume, cu ft = (644 ft) (3.5 ft) (5.0 ft) = 11,270 cu ft
    Next convert cubic feet to cubic yards. 1 cu yd = 27 cu ft;

    $\dfrac{11{,}270 \text{ cu ft}}{27 \text{ cu ft/cu yd}} = 417.41$ cu yd, round to 420

    Reference: AWWA, *Principles and Practices of Water Supply Operations, Basic Science Concepts and Applications,* Second Edition, Page 21, 86, and 107.

WATER DISTRIBUTION   223

14. Answer: **b.** 7.5

   Solution: First add all seven measurements.
   7.45 + 7.49 + 7.56 + 7.63 + 7.60 + 7.54 + 7.52 + 7.41 = 60.2

   $$\text{average} = \frac{\text{measurement sum}}{\text{no. of measurements}}$$

   $$= \frac{60.2}{8} = 7.525 \text{ pH, round to } 7.5 \text{ pH}$$

15. Answer: **b.** 45.993%

   Solution: $\frac{(34{,}411)(100\%)}{74{,}818} = 45.993\%$

   Reference: AWWA, *Principles and Practices of Water Supply Operations, Basic Science Concepts and Applications,* Second Edition, Page 62.

16. Answer: **d.** 635,000

   Solution: A 1% solution has 10,000 mg/L, so a 63.5% solution will have:

   $$\frac{(63.5\%)(10{,}000 \text{ mg/L})}{1\%} = 635{,}000 \text{ mg/L hypochlorite}$$

   AWWA, *Principles and Practices of Water Supply Operations, Basic Science Concepts and Applications,* Second Edition, Page 476.

17. Answer: **b.** 130 ft

   Solution: circumference = (π) (diameter)
   Rearrange the equation to solve for the diameter.

   $$\text{diameter} = \frac{\text{circumference}}{(\pi)}$$

   $$\text{diameter} = \frac{408.2 \text{ ft}}{3.14} = 130 \text{ ft}$$

   Reference: AWWA, *Principles and Practices of Water Supply Operations, Basic Science Concepts and Applications,* Second Edition, Page 72.

18. Answer: **a.** Amount of hydrogen ions released

   Reference: AWWA, *Principles and Practices of Water Supply Operations, Basic Science Concepts and Applications,* Second Edition, Page 423.

19. Answer: **a.** Coliform bacteria

   Reference: AWWA, *Water Distribution Operator Training Handbook,* Second Edition, Page 23, 28, and 29.

20. Answer: **a.** Plastic

    Reference: AWWA, *Water Distribution Operator Training Handbook*, Second Edition, Page 50.

21. Answer: **c.** 24 hours

    Reference: AWWA, *Principles and Practices of Water Supply Operations, Water Transmission and Distribution*, Second Edition, Page 206.

22. Answer: **b.** Hydraulically applied concrete-lined

    Reference: AWWA, *Water Distribution Operator Training Handbook*, Second Edition, Page 171.

23. Answer: **a.** Altitude

    Reference: AWWA, *Principles and Practices of Water Supply Operations, Water Transmission and Distribution*, Second Edition, Page 57.

24. Answer: **d.** So that small sections of water main may be shut off for maintenance

    Reference: AWWA, *Water Distribution Operator Training Handbook*, Second Edition, Page 60.

25. Answer: **d.** 43,567 sq ft

    Solution: First, find the surface area of the tank's wall. Imagine cutting the wall and rolling it out so that it is flat. The length of this now flat wall is simply the circumference of the tank, and is defined as: (diameter) (π). The area of the tank wall then is the length of the wall times the height.
    wall area, sq ft = (diameter) (π) (height)
    wall area, sq ft = (125.0 ft) (3.14) (48.5 ft) = 19,036 sq ft
    Next, find the top and bottom surface area.
    top and bottom area, sq ft = (2, top and bottom) (0.785) (D)^2 = (2) (0.785) (125.0 ft) (125.0 ft) = 24,531 sq ft
    total area of tank, sq ft = 19,036 sq ft + 24,531 sq ft = 43,567 sq ft

    Reference: AWWA, *Principles and Practices of Water Supply Operations, Basic Science Concepts and Applications*, Second Edition, Page 70 and 80.

26. Answer: **c.** 25,000 cu ft

    Solution: First, find the volume of soil removed from the trench in cubic feet.
    volume, cu ft = (L) (W) (D)
    volume, cu ft = (1,287 ft) (4.2 ft) (5.4 ft) = 29,189 cu ft,
    Next, convert 24 in. to feet.
    24 in./12 in. per foot = 2.0 ft

Next, determine the volume that the pipe occupies.

volume, cu ft = (0.785) (D)^2 (length)

volume, cu ft = (0.785) (2.0 ft) (2.0 ft) (1,287 ft) = 4,041 cu ft

Last, subtract the volume of the pipe from the volume of the trench.

volume of soil put in trench = 29,189 cu ft − 4,041 cu ft = 25,148 cu ft, rounded to 25,000

Reference: AWWA, *Principles and Practices of Water Supply Operations, Basic Science Concepts and Applications,* Second Edition, Page 86 and 87.

27. Answer: **d.** 5

   Reference: AWWA, *Water Distribution Operator Training Handbook,* Second Edition, Page 28.

28. Answer: **b.** 50 mg/L

   Reference: AWWA, *Water Distribution Operator Training Handbook,* Second Edition, Page 93.

29. Answer: **b.** Taste and odor

   Reference: AWWA, *Water Distribution Operator Training Handbook,* Second Edition, Page 34.

30. Answer: **b.** Silica sand. Portland cement and asbestos fibers are also used in asbestos-cement pipes.

   Reference: AWWA, *Water Distribution Operator Training Handbook,* Second Edition, Page 47.

31. Answer: **c.** Prevent surface water contamination from entering the water that is being drawn into the well

   Reference: AWWA, *Water Distribution Operator Training Handbook,* Second Edition, Page 187.

32. Answer: **a.** Mechanical joint

   Reference: AWWA, *Water Distribution Operator Training Handbook,* Second Edition, Page 60.

33. Answer: **a.** Cast-in-place concrete

   Reference: AWWA, *Water Distribution Operator Training Handbook,* Second Edition, Page 170.

34. Answer: **b.** 30.5 ft

   Solution: pressure, psi = $\dfrac{\text{depth of water}}{2.31 \text{ ft/psi}}$

Rearrange the equation to solve for depth.

depth of water = (pressure, psi) (2.31 ft/psi)

depth of water = (13.2 psi) (2.31 ft/psi) = 30.5 ft

Reference: AWWA, *Water Distribution Operator Training Handbook*, Second Edition, Page 248.

35. Answer: **b.** 6 hours

AWWA, *Water Distribution Operator Training Handbook*, Second Edition, Page 19.

36. Answer: **b.** Dissolved oxygen

Reference: AWWA, *Principles and Practices of Water Supply Operations, Water Treatment*, Second Edition, Page 268.

37. Answer: **d.** 300 mg/L

Reference: AWWA, *Water Distribution Operator Training Handbook*, Second Edition, Page 94

38. Answer: **c.** Under pressure

Reference: AWWA, *Water Distribution Operator Training Handbook*, Second Edition, Page 182.

39. Answer: **a.** Galvanized iron

Reference: AWWA, *Water Distribution Operator Training Handbook*, Second Edition, Page 124.

40. Answer: **a.** Night, when noise is lowest

Reference: AWWA, *Water Distribution Operator Training Handbook*, Second Edition, Page 118.

41. Answer: **b.** Ductile iron

Reference: AWWA, *Water Distribution Operator Training Handbook*, Second Edition, Page 45.

42. Answer: **d.** Year

Reference: AWWA, *Water Distribution Operator Training Handbook*, Second Edition, Page 36.

43. Answer: **a.** Distance from a well that the cone of depression affects the normal water level

Reference: AWWA, *Water Distribution Operator Training Handbook*, Second Edition, Page 185.

44. Answer: **a.** 0.30 mg/L

    Reference: *AWWA, Principles and Practices of Water Supply Operations, Water Quality,* Second Edition, Page 13, and Title 40 CFR 143.3.

45. Answer: **b.** Riser

    Reference: *AWWA, Water Distribution Operator Training Handbook*, Second Edition, Page 174.

46. Answer: **c.** Push-on joint

    Reference: *AWWA, Water Distribution Operator Training Handbook*, Second Edition, Page 44.

47. Answer: **b.** Connections to the curb stop

    Reference: *AWWA, Water Distribution Operator Training Handbook*, Second Edition, Page 136.

48. Answer: **b.** Change in water elevation from the normal level to the pumping level

    Reference: *AWWA, Water Distribution Operator Training Handbook*, Second Edition, Page 185.

49. Answer: **c.** Minneapolis style

    Reference: *AWWA, Water Distribution Operator Training Handbook*, Second Edition, Page 131.

50. Answer: **a.** Arch style

    Reference: *AWWA, Water Distribution Operator Training Handbook*, Second Edition, Page 131.

51. Answer: **c.** Piezometric surface

    Reference: *AWWA, Water Distribution Operator Training Handbook*, Second Edition, Page 182.

52. Answer: **a.** System size

    Reference: *AWWA, Principles and Practices of Water Supply Operations, Water Quality,* Second Edition, Page 24, and Title 40 CFR 141.86 (c).

53. Answer: **c.** Float mechanism

    Reference: *AWWA, Principles and Practices of Water Supply Operations, Water Transmission and Distribution,* Second Edition, Page 437.

54. Answer: **d.** Velocity meters

    Reference: *AWWA, Water Distribution Operator Training Handbook*, Second Edition, Page 145.

55. Answer: **a.** Velocity type

    Reference: *AWWA, Water Distribution Operator Training Handbook*, Second Edition, Page 145.

56. Answer: **a.** Magnetic

    Reference: *AWWA, Water Distribution Operator Training Handbook*, Second Edition, Page 147.

57. Answer: **a.** Proportional

    Reference: *AWWA, Water Distribution Operator Training Handbook*, Second Edition, Page 146.

58. Answer: **b.** Flush

    Reference: *AWWA, Water Distribution Operator Training Handbook*, Second Edition, Page 100 and 101, Figure 10-8.

59. Answer: **b.** Velocity

    Reference: *AWWA, Water Distribution Operator Training Handbook*, Second Edition, Page 192.

60. Answer: **c.** Freezing

    Reference: *AWWA, Water Distribution Operator Training Handbook*, Second Edition, Page 99.

61. Answer: **b.** 15 to 25 ft

    Reference: *AWWA, Water Distribution Operator Training Handbook*, Second Edition, Page 194.

62. Answer: **b.** Do last few turns manually

    Reference: *AWWA, Water Distribution Operator Training Handbook*, Second Edition, Page 67.

63. Answer: **c.** Wet-barrel

    Reference: *AWWA, Water Distribution Operator Training Handbook*, Second Edition, Page 100.

64. Answer: **b.** Centrifugal

    Reference: *AWWA, Water Distribution Operator Training Handbook*, Second Edition, Page 192.

65. Answer: **b.** Secondary instruments

    Reference: *AWWA, Water Distribution Operator Training Handbook*, Second Edition, Page 219.

66. Answer: **a.** Pressure-relief

Reference: AWWA, *Water Distribution Operator Training Handbook*, Second Edition, Page 64.

67. Answer: **b.** Warm-climate

Reference: AWWA, *Water Distribution Operator Training Handbook*, Second Edition, Page 100.

68. Answer: **b.** 69 mph

Solution: First, convert million gallons per day to gallons per minute.

gpm = (2.54 mgd)(1,000,000/1 M)(1 day/1,440 min) = 1,764 gpm, round to 1,760 gpm

$$\text{mhp} = \frac{(\text{flow, gpm})(\text{TH, ft})(8.34 \text{ lb/gal})}{(33,000 \text{ ft-lb/min/HP})(\text{motor efficiency})(\text{pump efficiency})}$$

Change percent motor and pump efficiencies to decimal form by dividing by 100%.

Motor efficiency: 87%/100% = 0.87

Pump efficiency: 79%/100% = 0.79

$$\text{mhp} = \frac{(1,760 \text{ gpm})(107 \text{ ft})(8.34 \text{ lb/gal})}{(33,000 \text{ ft-lb/min/HP})(0.87)(0.79)} = 69 \text{ mhp}$$

Reference: AWWA, *Principles and Practices of Water Supply Operations, Basic Science Concepts and Applications,* Second Edition, Page 299— Example 18 and Page 301— Example 20.

69. Answer: **a.** Globe

Reference: AWWA, *Water Distribution Operator Training Handbook*, Second Edition, Page 62.

70. Answer: **a.** Split-phase

Reference: AWWA, *Water Distribution Operator Training Handbook*, Second Edition, Page 203.

71. Answer: **b.** Vertical turbine

Reference: AWWA, *Water Distribution Operator Training Handbook*, Second Edition, Page 194.

72. Answer: **d.** Wound-rotor

Reference: AWWA, *Water Distribution Operator Training Handbook*, Second Edition, Page 204.

73. Answer: **a.** Globe

    Reference: *AWWA, Water Distribution Operator Training Handbook*, Second Edition, Page 62.

74. Answer: **c.** Thermocouple

    Reference: *AWWA, Water Distribution Operator Training Handbook*, Second Edition, Page 223.

75. Answer: **c.** 143 ft

    Solution:

    pressure head, ft = (pressure, psi) (2.31 ft/psi)

    pressure head, ft = (62 psi) (2.31 ft/psi) = 143 ft

    Reference: AWWA, Basic Science Concepts and Applications, Page 249, and *AWWA, Water Distribution Operator Training Handbook*, Second Edition, Page 248.

76. Answer: **b.** 6.7°F

    Solution:

    The formula for converting degrees Celsius to degrees Fahrenheit is:
    °F = 9/5 °C + 32°F

    °F = 9/5 (12.6°C) + 32°F = 54.68°F

    °F = 9/5 (16.3°C) + 32°F = 61.34°F

    61.34 °F − 54.68°F = 6.67°F

    Reference: *AWWA, Principles and Practices of Water Supply Operations, Basic Science Concepts and Applications,* Second Edition, Page 120, and *AWWA, Water Distribution Operator Training Handbook*, Second Edition, Page 255.

77. Answer: **c.** Ultrasonic meter. The Doppler effect is the apparent change in frequency of sound waves due to the relative velocity between the source of the sound waves and the observer.

    Reference: *AWWA, Water Distribution Operator Training Handbook*, Second Edition, Page 147.

78. Answer: **a.** Ohms

    Reference: *AWWA, Principles and Practices of Water Supply Operations, Basic Science Concepts and Applications,* Second Edition, Page 541.

79. Answer: **c.** Ultrasonic level-sensing system

    Reference: *AWWA, Principles and Practices of Water Supply Operations, Water Treatment,* Second Edition, Page 459, Figure 16-11.

80. Answer: **c.** Pump is losing efficiency

    Reference: AWWA, *Water Distribution Operator Training Handbook*, Second Edition, Page 191.

81. Answer: **d.** 280 ft

    Solution: pressure head, ft = (pressure, psi) (2.31 ft/psi)

    pressure head, ft = (121 psi) (2.31 ft/psi) = 279.51 ft, round to 280 ft

    Reference: AWWA, *Water Distribution Operator Training Handbook*, Second Edition, Page 248.

82. Answer: **a.** Transducer. Ultrasonic signals are also used to measure water depths in storage facilities.

    Reference: AWWA, *Water Distribution Operator Training Handbook*, Second Edition, Page 221.

83. Answer: **c.** 15 psi

    Reference: AWWA, *Water Distribution Operator Training Handbook*, Second Edition, Page 141, Table 13-1.

84. Answer: **a.** Pitot meter

    Reference: AWWA, *Principles and Practices of Water Supply Operations, Water Transmission and Distribution,* Second Edition, Page 434.

85. Answer: **c.** Decomposition of a material caused by an outside electric current

    Reference: AWWA, *Principles and Practices of Water Supply Operations, Water Transmission and Distribution,* Second Edition, Page 248.

86. Answer: **b.** 5 fps

    Reference: AWWA, *Water Distribution Operator Training Handbook*, Second Edition, Page 37.

87. Answer: **c.** Thermistor

    Reference: AWWA, *Water Distribution Operator Training Handbook*, Second Edition, Page 223.

88. Answer: **c.** 10 years

    Reference: AWWA, *Water Distribution Operator Training Handbook*, Second Edition, Page 153.

89. Answer: **b.** Class B

    Reference: No reference source specified

90. Answer: **b.** Methemoglobinemia

    Reference: *AWWA, Principles and Practices of Water Supply Operations, Water Quality,* Second Edition, Page 205.

91. Answer: **c.** 1.0 mg/L

    Reference: *AWWA, Principles and Practices of Water Supply Operations, Water Quality,* Second Edition, Page 13.

92. Answer: **d.** 0.05 mg/L

    Reference: *AWWA, Principles and Practices of Water Supply Operations, Water Quality,* Second Edition, Page 13.

93. Answer: **b.** National Primary Drinking Water Regulations

    Reference: *California State University, Small Water System Operation and Maintenance,* Third Edition, Chapter 2, Section 2.3 and Water Words.

94. Answer: **d.** Remove dangerous gases

    Reference: *California State University, Small Water System Operation and Maintenance,* Third Edition, Chapter 6, Section 6.20.

95. Answer: **d.** Class D

    Reference: *California State University, Small Water System Operation and Maintenance,* Third Edition, Chapter 7, Section 7.162.

96. Answer: **c.** Turbidity

    Reference: *AWWA, Principles and Practices of Water Supply Operations, Water Quality,* Second Edition, Chapter 3, Page 98.

97. Answer: **c.** Water from flowing in two directions

    Reference: *AWWA, Principles and Practices of Water Supply Operations, Water Transmission and Distribution,* Second Edition, Chapter 3, Page 74.

98. Answer: **c.** 90–180 degrees

    Reference: AWWA, Principles and Practices of Water Supply Operations, Water Transmission and Distribution, Second Edition, Appendix E, Page 565.

99. Answer: **d.** Water elevation

    Reference: *AWWA, Principles and Practices of Water Supply Operations, Basic Science Concepts and Applications,* Second Edition, Hydraulics 4, Page 246.

100. Answer: **b.** Reduce pressure across both sides to ease opening and closing

    Reference: *AWWA, Principles and Practices of Water Supply Operations, Water Transmission and Distribution*, Second Edition, Chapter 3, Page 65.

MATH FOR MORE PRACTICE, Answers

1. Answer: **b.** 39,626

    Solution: average no. of gallons per day $= \dfrac{\text{no. of gallons pumped}}{\text{no. of days}}$

    $= \dfrac{1{,}228{,}400 \text{ gal}}{31 \text{ days}} = 39{,}625.8$ gpd, round to 39,626 gpd

2. Answer: **a.** 0.8167

    Solution: $\dfrac{81.67\%}{100\%} = 0.8167$

3. Answer: **c.** 11.8%

    Solution: $(0.118)(100\%) = 11.8\%$

4. Answer: **a.** 2.7%

    Solution: % Lime $= \dfrac{(12 \text{ lb})(100\%)}{12 \text{ lb} + (8.34 \text{ lb/gal})(52 \text{ gal})}$

    $= \dfrac{(12 \text{ lb})(100\%)}{12 \text{ lb} + 433.68 \text{ lb}} = \dfrac{(12 \text{ lb})(100\%)}{445.68 \text{ lb}} = 2.7\%$ slurry

5. Answer: **d.** 70.9%

    Solution: percent $= \dfrac{(15{,}325)(100\%)}{21{,}611} = 70.9\%$

6. Answer: **a.** $x = 40$

    Solution: Write the proportion in fraction form and solve for x.

    $\dfrac{6}{72} = \dfrac{(x)}{480}$

    $x = \dfrac{(6)(480)}{72} = 40$ Thus, the proportion is: $6:72 = 40:480$.

7. Answer: **b.** 43 lb/day

    Solution: Set up a ratio and solve for the unknown, x.

    $\dfrac{x \text{ lb/day}}{12 \text{ cfs}} = \dfrac{25 \text{ lb/day}}{7 \text{ cfs}}$

    $x = \dfrac{(25 \text{ lb/day})(12 \text{ cfs})}{7 \text{ cfs}} = 42.86$ lb/day, round to 43 lb/day

8. Answer: **d.** 15,300 gpm

   Solution: no. of gallons per minute = (34 cfs) (60 sec/min) (7.5 gal/cu ft) = 15,300 gpm

9. Answer: **d.** 58,000 mg/L

   Solution: A 1% solution = 10,000 mg/L

   A 5.8% solution will have:

   $$\frac{(5.8\%)(10{,}000 \text{ mg/L})}{1\%} = 58{,}000 \text{ mg/L hypochlorite}$$

10. Answer: **b.** 236 ft

    Solution: circumference of circle = (π) (diameter) = (3.14) (75.0 ft) = 235.5 ft, round to 236 ft

11. Answer: **c.** 14.3 psi

    Solution: $\text{psi} = \dfrac{\text{depth, ft}}{2.31 \text{ ft/psi}} = \dfrac{33.11 \text{ ft}}{2.31 \text{ ft/psi}} = 14.3 \text{ psi}$

12. Answer: **b.** 84.4 psi

    Solution: pressure head, ft = (pressure, psi) (2.31 ft/psi)
    Rearrange the equation to solve for pressure.

    $\text{pressure, psi} = \dfrac{\text{pressure head, ft}}{2.31 \text{ ft/psi}} = \dfrac{195 \text{ ft}}{2.31 \text{ ft/psi}} = 84.4 \text{ psi}$

13. Answer: **d.** 46,032 sq ft

    Solution: no. of square feet = (length) (width) = (672 ft) (68.5 ft) = 46,032 sq ft

14. Answer: **a.** 119 sq in.

    Solution: area of triangle =

    $\dfrac{(\text{base})(\text{height})}{2} = \dfrac{(14 \text{ in.})(17 \text{ in.})}{2} = 119 \text{ sq in.}$

15. Answer: **c.** 7,623 cu ft

    Solution: volume, cu ft = (length) (width) (depth) = (363 ft) (3.5 ft) (6.0 ft) = 7,623 cu ft

WATER DISTRIBUTION 235

16. Answer: **b.** 350 cu yd

   Solution: First, find the number of cubic feet.
   volume, cu ft = (length) (width) (depth)
   volume, cu ft = (430 ft) (4.0 ft) (5.5 ft) = 9,460 cu ft
   Next convert cubic feet to cubic yards.

   1 cu yd = 27 cu ft, thus: $\dfrac{9{,}460 \text{ cu ft}}{27 \text{ cu ft/cu yd}}$ = 350 cu yd

17. Answer: **d.** 2,861 gal

   Solution: First, convert the diameter from inches to feet.

   no. of feet = $\dfrac{18 \text{ in.}}{12 \text{ in./ft}}$ = 1.5 ft

   Next, calculate the volume using the formula:
   no. of gallons = (0.785) (D^2) (length) (7.5 gal/cu ft)
   no. of gallons = (0.785) (1.5 ft) (1.5 ft) (216 ft) (7.5 gal/cu ft) = 2,861 gal

18. Answer: **d.** 16,179 cu ft

   Solution: First find the volume of the cone in cubic feet.
   volume, cu ft = ⅓πr^2 (depth); where the radius =
   diameter/2 = 21.0 ft/2 = 10.5 ft
   volume, cu ft = ⅓ (3.14) (10.5 ft) (10.5 ft) (14.2 ft) = 1,639 cu ft
   Next find the volume of the cylindrical part of the tank.
   volume = πr^2(depth) = (3.14) (10.5 ft) (10.5 ft) (42.0 ft) = 14,540 cu ft
   Then, add the two volumes for the answer.
   total volume, cu ft = 1,639 cu ft + 14,540 cu ft = 16,179 cu ft

19. Answer: **b.** 30.7 lb/day

   Solution: average amount of chlorine used, lb/day =

   $\dfrac{\text{sum of Cl}_2 \text{ used each day, lb}}{\text{total days}}$

   = $\dfrac{34 + 28 + 26 + 29 + 31 + 32 + 35}{7 \text{ days}}$ = 30.7 lb/day

20. Answer: **c.** 7.0 mgd

   Solution: average mgd flow = $\dfrac{\text{sum of mgd used each day}}{\text{total days}}$

   average mgd flow = $\dfrac{6.7 + 6.4 + 6.8 + 7.5 + 7.7 + 7.3 + 6.8}{7 \text{ days}}$

   = 7.03 mgd, round to 7.03 mgd

21. Answer: **d.** 2.2 fps

    Solution: First, convert the number of gallons per minute to cubic feet per second.

    $$\frac{550 \text{ gpm}}{(7.5 \text{ gal/cu ft})(60 \text{ sec/min})} = 1.2 \text{ cfs}$$

    Next, convert the diameter from inches to ft. (10 in.) (1 ft/12 in.) = 0.83 ft

    flow, cfs = (area, sq ft) (velocity, fps); where the area = (0.785) (D)^2

    1.2 cfs = (0.785) (0.83) (0.83) (flow, fps)

    Rearrange and solve for the flow in feet per second.

    $$\text{flow, fps} = \frac{1.2 \text{ cfs}}{(0.785)(0.83)(0.83)} = 2.2 \text{ fps}$$

22. Answer: **d.** 968 gpm

    Solution: flow, gpm = (flow, cfs) (7.5 gal/cu ft) (60 sec/min)
    = (2.15 cfs) (7.5 gal/cu ft) (60 sec/min) = 967.5 gpm, round to 968

23. Answer: **c.** 10.4 hr

    Solution: First, covert gallons per minute to gallons per hour.
    (4,810 gpm) (60 min/hr) = 288,600 gph

    Next, convert million gallons to gallons.
    (3.0 mil gal) (1,000,000/1 M) = 3,000,000 gal

    time, hours = 3,000,000 gal/288,600 gph = 10.4 hours

24. Answer: **d.** 38 cfs

    Solution: Q = VA or flow, cfs = (velocity, fps) (area, sq ft)
    Q, cfs = (2.10 fps) (6.5 ft) (2.8 ft) = 38.22 cfs, round to 38 cfs

25. Answer: **a.** 8.96 hr

    Solution: First convert million gallons to gallons.
    (2.00 mil gal) (1,000,000/1 M) = 2,000,000 gal

    and: (5.36 mgd) (1,000,000/1 M) = 5,360,000 gpd

    detention time, hours =

    $$\frac{(\text{tank volume})(24 \text{ hr/day})}{\text{flow, gpd}} = \frac{(2,000,000 \text{ gal})(24 \text{ hr/day})}{5,360,000 \text{ gpd}} = 8.96 \text{ hours}$$

WATER DISTRIBUTION 237

26. Answer: **a.** 6.65 hr

    Solution: First, convert pipe diameter in inches to feet.
    (12 in.) (1 ft/12 in.) = 1.0 ft
    Next, find the number of gallons in the pipeline.
    = (0.785) (D^2) (length, ft) (7.5 gal/cu ft)
    = (0.785) (1.0 ft) (1.0 ft) (2,485 ft) (7.5 gal/cu ft) = 14,630 gal
    Then convert million gallons per day to gallons per day.
    = (6.82 mgd) (1,000,000 gal/1 M) = 6,820,000 gpd
    Add the pipe and tank volume to get the total number of gallons.
    = 14,630 gal + 1,875,000 gal = 1,889,630 gal
    Using the following equation, solve for the detention time.

    $$\text{detention time, hours} = \frac{(\text{total volume})(24 \text{ hr/day})}{\text{flow, gpd}}$$

    $$\text{detention time, hours} = \frac{(1,889,630 \text{ gal})(24 \text{ hr/day})}{6,820,000 \text{ gpd}} = 6.65 \text{ hours}$$

27. Answer: **c.** 26 minutes

    Solution: First, convert the pipe diameter from inches to feet.
    (8 in.) (1 ft/12 in.) = 0.67 ft
    Next, determine the volume, in gallons, for 250 ft of the 8-in. pipeline.
    volume, gal = (0.785) (D^2) (length) (7.5 gal/cu ft)
    = (0.785) (0.67) (0.67) (250 ft) (7.5 gal/cu ft) = 660.7 gal, round to 661 gal
    Then, find the flushing time.

    $$\text{flushing time, min} = \frac{661 \text{ gal}}{25 \text{ gpm}} = 26.4 \text{ min, round to 26 min}$$

28. Answer: **b.** 1.55 mg/L

    Solution: chlorine demand, mg/L = chlorine dose, mg/L − chlorine residual, mg/L
    = chlorine demand, mg/L = 3.00 mg/L − 1.45 mg/L = 1.55 mg/L

29. Answer: **a.** 0.82 mg/L

    Solution: chlorine residual, mg/L = chlorine dose, mg/L − chlorine demand, mg/L
    = chlorine residual, mg/L = 2.75 mg/L − 1.93 mg/L = 0.82 mg/L

30. Answer: **d.** 198 lb/day

    Solution: lb/day = (mgd) (dosage, mg/L) (8.34 lb/day)
    = (8.65 mgd) (2.75 mg/L) (8.34 lb/gal) = 198.4 lb/day, rounded to 198

31. Answer: **d.** 1,131 lb/day

   Solution: lb/day = (mgd) (dosage, mg/L) (8.34 lb/day)
   = (38.75 mgd) (3.50 mg/L) (8.34 lb/gal) = 1,131 lb/day

32. Answer: **b.** 34 lb/day

   Solution: First, convert the pumping rate to million gallons per day.

   $$\frac{(\text{pumping rate, gpm})(1,440 \text{ min/day})}{1,000,000/M}$$

   $$= \frac{(975 \text{ gpm})(1,440 \text{ min/day})}{1,000,000/1M} = 1.40 \text{ mgd}$$

   Next, use the pounds equation to solve the problem.
   chlorine, lb/day = (mgd) (dosage, mg/L) (8.34 lb/gal)
   = (1.40 mgd) (2.90 mg/L) (8.34 lb/gal) = 33.86 lb/day,
   round to 34 lb/day

33. Answer: **a.** 0.73 lb

   Solution: First, convert gallons to million gallons.
   (1,500 gal) (1 M/1,000,000) = 0.0015 mil gal
   Then, determine the pounds of chlorine needed.
   no. lb, Cl_2 = (mil gal) (dosage, mg/L) (8.34 lb/gal)
   = (0.0015 mil gal) (35 mg/L) (8.34 lb/gal) = 0.44 lb of Cl_2
   0.44 / 0.605 = 0.73 lb of calcium hypochlorite

34. Answer: **b.** 1.83 gal

   Solution: First, find the volume of the pipe.
   volume, gal = (0.785) (D^2) (length) (7.5 gal/cu ft)
   = (0.785) (1.0) (1.0) (752 ft) (7.5 gal/cu ft) = 4,427.4 gal,
   round to 4,427 gal
   Next, find the number of million gallons.
   (4,427 gal) (1 M/1,000,000) = 0.0044 mil gal

   Then, use the pound equation.

   $$\text{lb of sodium hypochlorite} = \frac{\text{mil gal (dosage, mg/L)(8.34 lb/gal)}}{\% \text{ solution}/100\% \text{ available chlorine}}$$

   $$= \frac{(0.0044 \text{ mil gal})(25.0 \text{ mg/L})(8.34 \text{ lb/gal})}{6.00\%/100\% \text{ available chlorine}} = 15.3 \text{ lb}$$

   Last, calculate the number of gallons of sodium hypochlorite.
   15.3 lb/8.34 lb/gal = 1.83 gal

35. Answer: **b.** 200 gal

    Solution: First, determine the number of pounds of chlorine needed.
    lb = (mil gal) (dosage, mg/L) (8.34 lb/gal)
    = (2.5 mil gal) (50.0 mg/L) (8.34 lb/gal) = 1,042.5 lb, round to 1,040 lb
    Next, find the number of gallons of calcium hypochlorite.

    $$\text{calcium hypochlorite, gal} = \frac{(\text{chlorine, lb})(100\%)}{(8.34 \text{ lb/gal})(\% \text{ solution})}$$

    $$= \frac{(1{,}040 \text{ lb Cl}_2)(100\%)}{(8.34 \text{ lb/gal})(62.5\%)} = 199.5 \text{ gal, round to 200 gal}$$

36. Answer: **a.** 4.1 gal

    Solution: First, convert inches to feet.
    24 in./12 in. per foot = 2.0 ft
    Next, find the volume of the pipe.
    volume, gal = (0.785) (D^2) (length) (7.5 gal/cu ft)
    = (0.785) (2.0) (2.0) (750 ft) (7.5 gal/cu ft) = 17,663 gal
    Next, find the number of million gallons.
      (17,663 gal) (1 M/1,000,000) = 0.018 mil gal

    Then, use the pound equation.

    $$\text{lb of sodium hypochlorite} = \frac{(\text{mil gal})(\text{dosage, mg/L})(8.34 \text{ lb/gal})}{\% \text{ solution}/100\% \text{ available chlorine}}$$

    $$= \frac{(0.018 \text{ mil gal})(30.0 \text{ mg/L})(8.34 \text{ lb/gal})}{12.5\%/100\% \text{ available chlorine}} = 36 \text{ lb}$$

    Last, calculate the number of gallons of sodium hypochlorite.
    sodium hypochlorite, gal = 34 lb/8.34 lb/gal = 4.07 gal, round to 4.1 gal

37. Answer: **d.** 31 mg/L

    Solution: First, find the number of feet in 1.4 miles.
    (5,280 ft/mile) (1.4 miles) = 7,392 ft
    Next, find the volume in cubic feet for the pipe.
    volume, gal = (0.785) (D^2) (length, ft) (7.5 gal/cu ft)
      = (0.785) (2.0 ft) (2.0 ft) (7,392 ft) (7.5 gal/cu ft) = 174,081.6 gal,
      round to 174,082 gal

    Convert the number of gallons to million gallons.

    $$\frac{174{,}082 \text{ gal}}{1{,}000{,}000/1 \text{ M}} = 0.17 \text{ mil gal}$$

$$\text{dosage, mg/L} = \frac{\text{lb of chlorine}}{(\text{mil gal})(8.34 \text{ lb/gal})}$$

$$= \frac{44 \text{ lb}}{(0.17 \text{ mil gal})(8.34 \text{ lb/gal})} = 31.03 \text{ mg/L, round to 31 mg/L}$$

38. Answer: **b.** 6%

    Solution: %HTH solution = $\dfrac{(\text{lb HTH})(100\%)}{(\text{no. of gal})(8.34 \text{ lb/gal})}$

    $$= \frac{(125 \text{ lb HTH})(100)}{(250 \text{ gal})(8.34 \text{ lb/gal})} = 5.99\%, \text{ round to 6.0\% HTH solution}$$

39. Answer: **c.** 10 gpm/ft

    Solution: specific yield, gpm/ft = $\dfrac{\text{well yield, gpm}}{\text{drawdown, ft}}$

    $$= \frac{162 \text{ gpm}}{16 \text{ ft}} = 10.125 \text{ gpm/ft, round to 10 gpm/ft}$$

40. Answer: **d.** 22.7 ft

    Solution: drawdown, ft = $\dfrac{\text{well yield, gpm}}{\text{specific yield, gpm/ft}}$

    $$= \frac{265 \text{ gpm}}{11.7 \text{ gpm/ft}} = 22.65 \text{ ft, round to 22.7 ft}$$

41. Answer: **a.** 19.7 ft

    Solution: drawdown, ft = pumping water level, ft − static water level, ft = 104.2 ft − 84.5 ft = 19.7 ft

42. Answer: **d.** 105.2 ft

    Solution: drawdown, ft = pumping water level, ft − static water level, ft

    Rearrange the equation to solve for pumping water level.
    pumping water level, ft = drawdown, ft + static water level
    = 79.12 ft + 26.08 ft = 105.2 ft

43. Answer: **c.** 326,400 gal

    Solution: First, convert hours the pump worked to minutes.
    8 hr × 60 min/hr = 480 min
    Then, determine the number of gallons that it will discharge in 8 hours.
    no. of gal discharged = (no. of min) (pump rate, gpm)
    no. of gal discharged = (480 min) (680 gpm) = 326,400 gal

44. Answer: **a.** 0.22 gpm

    Solution: First, determine the number of gallons that was pumped.
    no. of gal = (0.785) (diameter^2) (level drop, ft) (7.5 gal/cu ft)
    = (0.785) (2.0 ft) (2.0 ft) (2.25 ft) (7.5 gal/cu ft) = 52.99 gal, round to 53 gal
    Next, find the number of minutes in 4 hours.
    (4 hr) (60 min/hr) = 240 min
    Last, divide the number of gallons pumped by the time in minutes.

    $$\text{no. of gpm} = \frac{53 \text{ gal}}{240 \text{ min}} = 0.22 \text{ gpm}$$

45. Answer: **c.** 22.7 ft

    Solution: total head, ft = total static head, ft + head losses, ft
    = 19 ft + 3.7 ft = 22.7 ft

46. Answer: **b.** 22 bhp

    Solution: brake hp = (hp) (motor efficiency)
    brake hp = (25 hp) (89%/100% motor efficiency) = 22 bhp

47. Answer: **d.** both column A and B = 38.0 psi

    Solution: 88 ft × 0.433 = approximately 38 psi

48. Answer: **b.** 448 kW

    Solution: First add the total horsepower.
    300 hp + 60 hp + 100 hp + 25 hp + 100 hp + 11 hp + 4 hp = 600 hp
    kW = (no. of hp) (0.746 kW/hp) = (600 hp) (0.746 kW/hp) = 448 kW

49. Answer: **c.** 0.017 mg/L

    Solution: First, line up the results in one column from largest at the top to smallest at the bottom.

    0.019 mg/L
    0.017 mg/L   90th percentile
    0.013 mg/L
    0.011 mg/L
    0.010 mg/L
    0.010 mg/L
    0.009 mg/L
    0.008 mg/L
    0.006 mg/L
    0.005 mg/L

    For convenience there are 10 analytical results, each equivalent to 10%. Count up from the bottom nine places (9 times 10% = 90% or the 90th percentile). The 90th percentile is 0.017 mg/L.

50. Answer: **d.** 417 lb

    Solution: %HTH solution = $\dfrac{(\text{lb HTH})(100\%)}{(\text{no. of gal})(8.34 \text{ lb/gal})}$

    no. lb of HTH = (% solution) (no. of gal) (8.34 lb/gal)/100%
    = (10.0% solution) (500 gal) (8.34 lb/gal)/100% = 417 lb of HTH

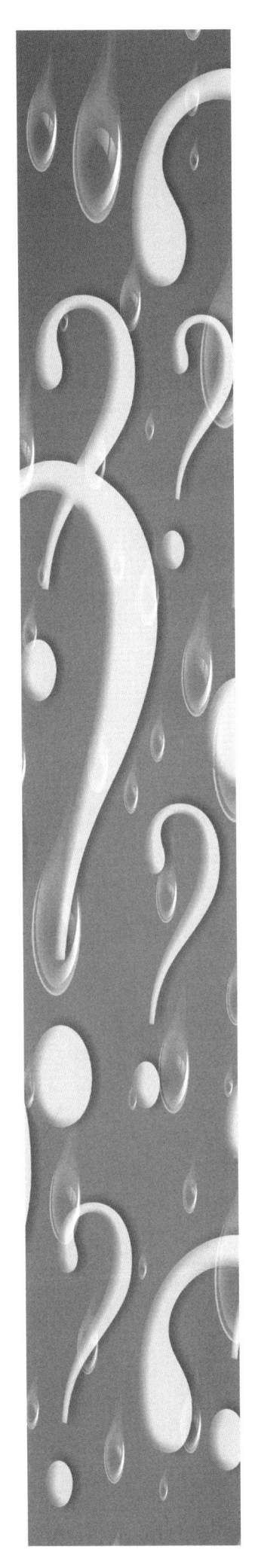

APPENDIX A

Formulas, Conversion Factors, and Abbreviations

Formulas/Conversion Tables

NOTE: This is the table that appears in ABC Certification Exams. Since ABC develops both water and wastewater examinations, this formula/conversion table includes both water and wastewater formulas.

$$\text{Acid Feed Rate} = \frac{(\text{Waste Flow})(\text{Waste Normality})}{\text{Acid Normality}}$$

$$\text{Alkalinity} = \frac{(\text{mL of Titrant})(\text{Acid Normality})(50{,}000)}{\text{mL of Sample}}$$

Area of Circle = $(0.785)(\text{Diameter}^2)$ or $(\pi)(\text{Radius}^2)$

Area of Cylinder = $[(0.785)(\text{Diameter} \times 2)] + [(\pi)(\text{Diameter})(\text{Height})]$

Area of Rectangle = (Length)(Width)

$$\text{Area of Triangle} = \frac{(\text{Base})(\text{Height})}{2}$$

$$\text{Chemical Feed Pump Setting, \% Stroke} = \frac{(\text{Desired Flow})(100\%)}{\text{Maximum Flow}}$$

Chemical Feed Pump Setting, mL/min =
$$\frac{(\text{Flow, mgd})(\text{Dose, mg/L})(3.785 \text{ L/gal})(1{,}000{,}000 \text{ gal/mil gal})}{\text{Maximum Flow}}$$

Circumference of Circle = (3.14)(Diameter)

Composite Sample Single Portion =
$$\frac{(\text{Instantaneous Flow})(\text{Total Sample Volume})}{(\text{Number of Portions})(\text{Average Flow})}$$

$$\text{Detention Time} = \frac{\text{Volume}}{\text{Flow}}$$

$$\text{Digested Sludge Remaining, \%} = \frac{(\text{Raw Dry Solids})(\text{Ash Solids})(100\%)}{(\text{Digested Dry Solids})(\text{Digested Ash Solids})}$$

$$\text{Discharge} = \frac{\text{Volume}}{\text{Time}}$$

Dosage, lb/day = (mg/L)(8.34)(mgd)

Davidson Pie Diagram

The Davidson pie diagram is a great method for visualizing how to set up a dosage problem. The diagram (shown below) is used by covering up the unknown while leaving the known uncovered. What remains is the equation that should be used to solve for the unknown.

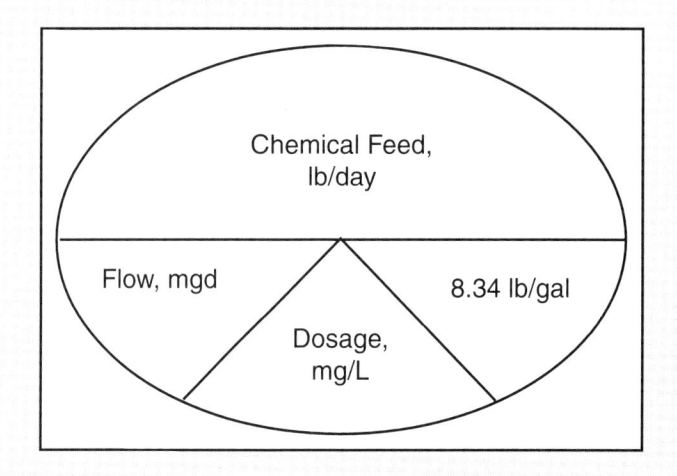

Permission: Davidson Pie Diagram, by Mr. Gerald Davidson, City of Merced (California).

For example, in the following equation the unknown quantity is flow, in million gallons per day. What is left in the diagram is chemical feed, in pounds per day, over flow, in dosage of milligrams per liter and 8.34 lb/gal. The equation is written as follows:

$$\text{Flow, mgd} = \frac{\text{Chemical Feed, lb/day}}{(\text{Dosage, mg/L})(8.34 \text{ lb/gal})}$$

$$\text{Efficiency, \%} = \frac{(\text{In} - \text{Out})(100\%)}{\text{In}}$$

$$\text{Feed Rate, lb/day} = \frac{(\text{Dosage, mg/L})(\text{Capacity, mgd})(8.34 \text{ lb/gal})}{(\text{Available Fluoride Ion})(\text{Purity})}$$

$$\text{Feed Rate, gal/min (Saturator)} = \frac{(\text{Plant Capacity, gpm})(\text{Dosage, mg/L})}{(18{,}000 \text{ mg/L})}$$

$$\text{Filter Backwash Rate} = \frac{\text{Flow}}{\text{Filter Area}}$$

$$\text{Filter Yield, lb/hr/sq ft} = \frac{(\text{Solids Loading, lb/day})(\text{Recovery, \%}/100\%)}{(\text{Filter Operation, hr/day})(\text{Area, ft}^2)}$$

APPENDIX A   247

$$\text{Food/Microorganism Ratio} = \frac{\text{BOD, lb/day}}{\text{MLVSS, lb}}$$

$$\text{Gallons/Capita/Day} = \frac{\text{Gallons/Day}}{\text{Population}}$$

$$\text{Hardness} = \frac{(\text{mL of Titrant})(1{,}000)}{\text{mL of Sample}}$$

$$\text{Horsepower} = \frac{(\text{Flow, gpm})(\text{Head, ft})}{(3{,}960)(\text{Efficiency})}$$

$$\text{Hydraulic Loading Rate} = \frac{\text{Flow}}{\text{Area}}$$

Mean Cell Residence Time (MCRT) =

$$\frac{\text{Suspended Solids in Aeration Systems, lb}}{\text{Suspended Solids Wasted, lb/day} + \text{Suspended Solids Lost, lb/day}}$$

$$\text{Organic Loading Rate} = \frac{\text{Organic Load, lb BOD/day}}{\text{Volume}}$$

$$\text{Oxygen Uptake} = \frac{\text{Oxygen Usage}}{\text{Time}}$$

$$\text{Population Equivalent} = \frac{(\text{Flow, mgd})(\text{BOD, mg/L})(8.34 \text{ lb/gal})}{\text{lb BOD/day/person}}$$

$$\text{Reduction in Flow, \%} = \frac{(\text{Original Flow} - \text{Reduced Flow})(100\%)}{\text{Original Flow}}$$

$$\text{Slope} = \frac{\text{Drop or Rise}}{\text{Distance}}$$

$$\text{Sludge Age} = \frac{\text{Mixed Liquor Solids, lb}}{\text{Primary Effluent Solids, lb/day}}$$

$$\text{Sludge Index} = \frac{\%\text{ Settleable Solids}}{\%\text{ Suspended Solids}}$$

$$\text{Sludge Volume Index} = \frac{(\text{Settleable Solids, \%})(10{,}000)}{\text{MLSS, mg/L}}$$

Solids Applied, lb/day = (Flow, mgd)(Concentration, mg/L)(8.34 lb/gal)

Solids Concentration = $\dfrac{\text{Weight}}{\text{Volume}}$

Solids Loading, lb/day/sq ft = $\dfrac{\text{Solids Applied, lb/day}}{\text{Surface Area, sq ft}}$

Solids, mg/L = $\dfrac{(\text{Dry Solids, grams})(1{,}000{,}000)}{\text{mL of Sample}}$

Surface Loading Rate = $\dfrac{\text{Flow}}{\text{Area}}$

Velocity = $\dfrac{\text{Flow}}{\text{Area}}$ or $\dfrac{\text{Distance}}{\text{Time}}$

Volatile Solids, % = $\dfrac{(\text{Dry Solids} - \text{Ash Solids})(100\%)}{\text{Dry Solids}}$

Volume of Cube = (Length) (Width) (Height)

Volume of Cone = (⅓) (0.785) (Diameter^2) (Height)

Volume of Cylinder = (0.785) (Diameter^2) (Height)

Waste Milliequivalent = (mL) (Normality)

Waste Normality = $\dfrac{(\text{Titrant Volume})(\text{Titrant Normality})}{\text{Sample Volume}}$

Weir Overflow Rate = $\dfrac{\text{Flow}}{\text{Weir Length}}$

Conversion Factors

acre	=	43,560 square feet
1 cubic foot	=	7.481 gallons
1 foot	=	0.305 meters
1 gallon	=	3.785 liters
1 gallon	=	8.34 pounds
1 grain per gallon	=	17.1 mg/L
1 horsepower	=	0.746 kilowatts
1 million gallons per day	=	694 gallons per minute
1 pound	=	0.454 kilograms
1 pound per square inch	=	2.31 feet of water
1%	=	10,000 mg/L
Degrees Celsius	=	(Degrees Fahrenheit − 32) ($5/9$)
Degrees Fahrenheit	=	(Degrees Celsius) ($9/5$) + 32

Abbreviations

%	percent
BOD	biochemical oxygen demand
cfm	cubic feet per minute
cfs	cubic feet per second
cu ft	cubic feet
DO	dissolved oxygen
fps	feet per second
ft	foot, feet
g	gram
gal	gallon
gpd	gallons per day
gpg	grains per gallon
gph	gallons per hour
gpm	gallons per minute
hp	horsepower
hr	hour
in.	inch
lb	pounds
mg/L	milligrams per liter
mgd	million gallons per day
mhp	motor horsepower
mL	milliliter
MLSS	mixed liquor suspended solids
MLVSS	mixed liquor volatile suspended solids
oz	ounce
ppm	part per million
psi	pounds per square inch
sq ft	square feet
whp	water horsepower
yr	year

ADDITIONAL RESOURCES

Water Treatment Operator Handbook

This book covers everything water treatment operators need to know to perform their jobs and keep in compliance with changing regulations. Every phase of a water treatment operator's job is addressed.

Edition: 2002, Softbound, 252 pp.
ISBN 1-58321-184-5; Catalog Number 20481.

Water Distribution Operator Training Handbook, Second Edition

A comprehensive text for water distribution operator training.

Edition: 1999, Softbound, 278 pp.

ISBN 1-58321-014-8; Catalog Number 20428.

Water Supply Operations I: Water Sources Textbook

Course textbook teaches the basics of the development of water sources for drinking water, groundwater, surface water, raw water characteristics, and the hydrologic cycle.

Edition: 2003, Hardback, 210 pp.

ISBN 1-58321-229-9; Catalog Number 1955.

Water Supply Operations II: Water Treatment Textbook

WSO Water Treatment provides a valuable introduction to the fundamentals of common water treatment processes and techniques.

Edition: 1995, Hardback, 523 pp.

ISBN 0-89867-789-0; Catalog Number 1956.

Water Supply Operations III: Water Transmission and Distribution Textbook

Get comprehensive coverage of the basic principles behind the design, construction, operation, and maintenance of water distribution systems in WSO Water Transmission and Distribution.

Edition: 1996, Hardback, 630 pp.

ISBN 0-89867-821-8; Catalog Number 1957.

Water Supply Operations IV: Water Quality Textbook

Contains information about methods of water quality analysis and drinking water regulations.

Edition: 1995, Hardback, 252 pp.

ISBN 0-89867-804-8; Catalog Number 1958.

Water Supply Operations V: Basic Science Concepts and Applications Textbook

Fnd the practical discussions you need of mathematics, hydraulics, chemistry, and electricity in WSO Basic Science.

Edition: 1995, Hardback, 670 pp.

ISBN 0-89867-796-3; Catalog Number 1959.

Basic Chemistry for Water and Wastewater Operators

A basic chemistry primer tailored for operators of drinking water or wastewater systems.

Edition: 2002, Softbound, 178 pp.

ISBN 1-58321-148-9; Catalog Number 20494.

Basic Microbiology for Drinking Water Personnel

Basic Microbiology for Drinking Water Personnel provides clear, short descriptions of the waterborne microorganisms—bacteria, viruses, protozoa, amoebae, and algae—which either pose a human health threat or contribute to distribution system corrosion.

Edition: 2001, Softbound, 85 pp.

ISBN 1-58321-121-7; Catalog Number 20463.

For pricing and ordering information, please visit the online bookstore at www.awwa.org/bookstore or call AWWA Customer Service at 1.800.926.7337.